Marek Kohn is the author of three other books, including *The Race Gallery*, hailed by the *Guardian* as 'an elegant, timely and devastating critique of racism in science', and *Dope Girls*. He has been a freelance journalist since 1983, and his 'Second Site' column appears in the *Independent on Sunday*.

As We Know It

Coming to Terms with
an Evolved Mind

MAREK KOHN

Granta Books
London

For Sue and Teo

Granta Publications, 2/3 Hanover Yard, London N1 8BE

First published in Great Britain by Granta Books 1999
This edition published by Granta Books 2000

A CIP catalogue record for this book is
available from the British Library.

1 3 5 7 9 10 8 6 4 2

ISBN 1 86207 368 6

Typeset by M Rules

Printed and bound in Great Britain by Mackays of Chatham plc

Contents

Acknowledgements   ix

1 Costs   1

2 Symmetry   47

3 Trust   169

4 Benefits   225

Notes and References   295
Further Reading   313
Index   316

Acknowledgements

Thanks to Helena Cronin, for introducing me to modern Darwinian theory, for rising to scepticism in good part, and for enabling me (and many others) to listen to the ideas of a wealth of inspiring and provocative scientists; to Rob Kruszynski, for long discussions and comprehensive references in his personal capacity, and for allowing me to hold the Furze Platt Giant in my own hands, as well as letting me see the Museum's trove of handaxes – more splendid in my eyes than jewels! – in his official capacity as a curator at the Natural History Museum, to which thanks is also due; to Chris Knight and Steve Mithen for papers, conversation and the opportunity to engage with their ideas at length; to Ian Watts, Richard Wilkinson, Kenan Malik, Rosalind Arden, Chris Stringer and Oliver Curry, for similar reasons; to Rob Foley, Geoff Miller and Camilla Power, for their valuable comments on a draft text; to Mark Roberts and his colleagues for allowing me to visit Boxgrove; to Phil Harding for knapping me a handaxe, one of my favourite pieces of silicon-based technology; to Karen Perkins at the Quaternary Section of the British Museum's Department of

Romano-British Antiquities, for showing me some of the Section's collection of handaxes (like the Natural History Museum's, but much, much bigger); to my editor Neil Belton, for his work on the book, his sympathetic approach to the difficulties of getting it written, and for his steadfast commitment to my writing, which has given me the ground beneath my feet that any author needs.

Above all, thanks to my wife Sue, without whose love, kindness and understanding I could not have undertaken this project; and thanks also to Patricia Arens and Julie Baldwin for looking after our son Teo, despite whom I finished it.

ONE

Costs

1

A million years ago there were minds that may have been human, but not as we know it. By that time, beings with minds like these had occupied much of the Old World, from the south of Africa to the east of Asia. Sometime between then and now, our ancestors made an infinitely more important migration, and moved into the imagined world of symbols, language, culture and magic.

They thus became human as we know it, and we are left with the question of how to understand this change of state. On the one hand, we are animals whose brains have evolved by the same processes as any other feature of living organisms. On the other, we now live within a sphere of artificial information in which changes can be made in pursuit of goals, instead of the single blind steps that natural evolution takes, and which in many other respects has a life of its own. The temptation is to see this as a higher realm, and a land of exemptions, where humans run free of the laws that govern all other species.

Humans are special, unarguably, and a partition between

biology and culture gives us a tidy scheme within which we can make sense of our lives. But, like many borders, it is largely a fiction, and sooner or later is likely to cause trouble. It also places a barrier in front of one of the most profound and marvellous questions we could possibly ask. How could beings without culture create it, and move into this imagined world? To explore this question, we need to contemplate what minds were like before culture, as best we can. We have to think of our ancestors in the hominid family as Darwinian creatures and nothing but, in order to work out why it was in these beings' reproductive interests to enter the imagined world, and to specify the conditions under which this transition could have been possible. In short, we need an evolutionary explanation for the origins of culture. Without that, we can't fully come to terms with an evolved mind.

Perhaps we aren't ready to do this, but we should; not just for the sake of truth and self-knowledge, but because of the things that are wrong with the way we live, and because we need to recover a confidence that we can make them better. *As We Know It* offers a Darwinian idea that may help make sense of minds a million years old, and describes a radical evolutionary theory of how language and culture began. They are presented not as pat solutions to great questions, but as illustrations of the kind of questions that we need to ask in order to know our minds.

As far as this book is concerned, questions about consciousness are not among them. It treats consciousness as a critical evolutionary issue in just one important respect, as the basis of theory of mind, the ability to think what another individual might be thinking. It does not attempt to explore the phenomena of consciousness, the redness of red and the wetness of water. Maybe one day these will be convincingly analysed in terms of natural selection, but for the foreseeable future they will remain matters of electricity and chemistry.

Seeing consciousness as the principal question of human

mental evolution is a way of keeping faith with the soul by secular means. But a mind in isolation is nothing. If one starts from the mystery of subjective consciousness, the questions are profound, but their relation to everyday life may be tenuous. Coming to terms with an evolved mind is challenging because of the questions it poses about our relations with each other; about how our evolved psychology guides us in our approaches to sex, status and fairness. *As We Know It* starts with four feet on the ground, and works from there. Hominids are treated as animals in a landscape, finding food and choosing mates, rather than as creatures of destiny. They are treated also as social animals, who have got to where they are today through the evolutionary motive power generated by interacting with each other.

In dealing with such issues, we are faced with the possibility that we know less than we would like to think we know about what is going on in our minds. Freud presented this problem in terms that merely deepened the sense of mystery, describing unconscious mental depths filled with currents of primordial instinct and ancient taboos. While agreeing with Freud that this is largely about sex, modern evolutionary theory depicts a mind fit for a systems analyst rather than a priest, its architecture built from modules and algorithms. To understand its implications, we have to consider the environment to which it is adapted. First and foremost, the environment of mind is society. For that reason, this book ends up not in the depths of the self, but in modern society and its politics.

In making a way to that destination, I have chosen not to follow some of the more familiar routes through human evolution. Instead of concentrating on the artistic achievements of the Upper Palaeolithic, traditionally used to show that people like us were so much more sophisticated than their predecessors, I have dwelt on artefacts made in the Lower Palaeolithic by hominids who had no symbolic culture. Considering these objects, known as handaxes, affords an opportunity to contemplate minds which were radically different from ours, and may or may not be considered human.

When imagining another mind, it is only natural to wonder what it must be like to be that individual. We would not be human if we didn't – according to the theory behind theory of mind. The means we have for imagining what it was like to be a Lower Palaeolithic hominid are indirect in the extreme. We have to build up a picture of how these ancestors lived using a combination of ecology and archaeology. Working from the ground up, we need to establish what we can about the landscape: its rocks, its watercourses, its vegetation, the sunlight and rain that fell upon it. These conditions shaped the forms that hominid societies took, and the possibilities can be explored by careful comparisons among other living primate species. Having created a context using general primate knowledge, we can try to make sense of handaxes, which are what remains of the hominids' distinctive character. It may be no more than human vanity, but in will-o'-the-wisps one can imagine moments of being a hominid a million years ago.

2

We do not migrate from biology to culture like settlers moving into a land of opportunity, nor do we shuttle backwards and forwards between the two provinces. We live in both at the same time. This is a complicated state of affairs. The options are to try to understand it better, to maintain the fiction of simplicity, or to declare that the situation is too complicated to analyse.

It is not just a matter of what is intellectually taxing, though, but of what is morally tainted. Many people cannot hear 'Darwinism' and 'society' in the same breath without thinking of Social Darwinism, the ideology of fitness that gave evolution a bad name. Its most extreme manifestation was in Nazism, and a widespread suspicion remains that any move to bring Darwinism into social affairs will be a step back in that direction. Social Darwinism is easier to recognize than to define precisely, but it is generally taken to incorporate belief in racial hierarchy, concern about the number of children born to people of low social status, and the assumption that success in life reflects innate quality rather than a fortunate position in an unfairly ordered society. Some of its themes are closer to prevailing wisdom than conventional opinion would care to admit.

Social Darwinism is therefore a phrase that reliably triggers moral revulsion, and modern Darwinism disclaims the label vehemently. The philosopher Daniel Dennett is not one to sugar the Darwinian pill as far as its implications for religion are concerned, but when it comes to politics, he feels obliged to denounce Social Darwinism as 'an odious misapplication of Darwinian thinking in defense of political doctrines that range from callous to heinous'.[1]

In his book, *Darwin's Dangerous Idea*, Dennett argues that there is widespread resistance to the Darwinian idea, taking a variety of forms. He concentrates not on those who take the Book of Genesis literally – even though they constitute 48 per cent of the US population, according to a poll he mentions – but on scholars who accept that living organisms are the modified descendants of earlier forms. Dennett relates the history of Darwinism as a series of strategic retreats by its opponents, in which they accept that Darwinism can explain a certain fraction of the world's great questions, but attempt to draw a line beyond which it cannot pass.

While Dennett focuses on individuals, the process he identifies is well illustrated by the response of the Catholic Church to the challenge of evolution. The Holy See came to terms with Darwinism quite swiftly by its standards. Pope Pius XII gave Catholics permission to accept evolutionary theory in 1950, but insisted that divine intervention had been necessary to create the human soul. In 1996, Pope John Paul II moved the Church's position from tolerance to endorsement, when he agreed that the theory of evolution is 'more than just a hypothesis'. While this accentuated the positive side of Catholic engagement with science, it did not alter the strategy of defining spheres of influence. Science was allowed the material domain, as long as it did not entertain pretensions to the spiritual realm. As the Pontiff made clear in later remarks, the Church considered evolution 'second-ary' to divine action. 'Evolution isn't enough to explain the origins of humanity,' he said, 'just as biological chance alone isn't enough to explain the birth of a baby.'[2] The policy could also be seen at work in a conference on the development of the universe, attended by the cosmologist Stephen Hawking, which the Pontifical Academy of Sciences held at the Vatican in 1981. John Paul II set the boundary at the Big Bang, advising the scientists present that they were free to go back as far as that point. Any fur-ther, and the physicists would have to hand over to the metaphysicians.[3]

Unlike the Church, with a political understanding of the need for historic compromise developed over two millennia, ordinary people tend not to be overly concerned with dogma. But you don't have to be a Catholic to share the feeling that something about humanity must be kept sacred. For some, this will be expressed in the idea of a spirit. Others will affirm that those aspects of the human condition which science cannot describe are the ones we should hold most precious. Faced with scientific pre-tensions to explain fundamental aspects of culture, they might well fear that science was mounting a bid to dehumanize human-ity. And once again, that raises the spectre of what happened when Social Darwinism was incorporated into political doctrine.

There is no alternative, though. To come to terms with an evolved mind, you have to learn to stop worrying and love socio-biology.

The word makes a lot of people wince, including many of the sci-ence's practitioners. It made its public début with the publication of Edward O. Wilson's *Sociobiology: The New Synthesis*, in 1975. Thanks to the final chapter, which discussed the human species, the book had notoriety thrust upon it. Critics identified sociobi-ology as the science department of the New Right, an association which has persisted ever since. 'Sociobiology is yet another

attempt to put a natural scientific foundation under Adam Smith,'
wrote Steven Rose, Richard Lewontin and Leon Kamin. 'It com-
bines vulgar Mendelism, vulgar Darwinism, and vulgar
reductionism in the service of the status quo.'[4]

For Rose, Lewontin and Kamin, Darwinian sociobiology was
no more than an updated edition of Victorian Social Darwinism.
The political critics of sociobiology saw it as an adjunct to the
aggressive ideology of the free market, justifying selfishness, com-
petitiveness, individualism and the survival of the fittest. Their
assaults left their mark on the public face of sociobiology, encouraging a second wave of sociobiological thinkers to distance
themselves from the pioneers.

These theorists call themselves evolutionary psychologists,
and represent what is now by far the most influential current in
sociobiological thought. Opinions vary on the difference
between sociobiology and evolutionary psychology. According
to Donald Symons, a notable exponent of the latter, human
sociobiology is one of several labels – he himself favours the
term 'Darwinian social science' – applied to a school of thought
based on the idea that individuals seek to 'maximize inclusive
fitness'. This means that they will always tend to behave in ways
which favour the reproduction of their own genes and those they
share with their kin.[5] They should therefore do everything possible to maximize the number of offspring who survive long
enough to reproduce in their turn, and Symons points out how
remiss people are in this respect. A healthy white woman wishing to maximize her fitness in the United States today could
bear a baby every year or two, he suggests, in full confidence that
each would be adopted by a middle-class family. 'A reasonably
young male member of the *Forbes* 400 could use his fortune to
construct a reproductive paradise in which women and their children could live in modest affluence and security for life (as long
as paternity was verified),' he continues. A wealthy man's repro-
ductivity could be increased a hundredfold in this way.

Meanwhile, men would vie with each other to make deposits in sperm banks: Symons's mischievous logic implies that they would compete more fiercely over these than they would over actual women. However, he concludes, 'It is difficult to picture clearly a modern industrial society in which people strive to maximize inclusive fitness because such a society would have so little in common with our own.'

Evolutionary psychology is less convinced that human genes know what's good for them these days. It sees living humans as creatures shaped by an ancestral environment long since left behind, which it calls the environment of evolutionary adaptation, or EEA. The idea was first raised in 1969 by the child psychologist John Bowlby, in the first volume of his work, *Attachment and Loss*. Evolutionists then took it over and rephrased it in the modern Darwinian terms of inclusive fitness, rather than the good of the group. In the first instance, the EEA was the savannah, and subsequently expanded to include most environments that the Earth's land surface has to offer. All these environments supported modes of life that were fundamentally similar, in that they were based on gathering and hunting. Farming only started within the last ten thousand years, a footnote to the chapter of *Homo* that had lasted a couple of million years. Evolutionary psychology conceives the mind as a collection of adaptations made to address specific problems in the founding environment. In contemporary environments, people's behaviour will sometimes be adaptive, and sometimes not.

There is no question that this is fundamental to evolutionary psychology's vision. Nevertheless, evolutionary psychology can be seen as a matured version of 1970s sociobiology, with a better sense of history. Many insiders certainly seem to feel that the choice of name is a political one. In 1996, the Human Behavior and Evolution Society officially dropped the word 'sociobiology'. One scholar, contributing to an e-mail discussion about the decision, revealed that although he used the term 'evolutionary

psychology' to describe his university classes, he privately
remained 'a reformed, unrepentant sociobiologist'.[6]

Standard definitions of sociobiology, derived from Edward
O. Wilson, refer to the study of the biological bases of social
behaviour. Under those terms, evolutionary psychology is a sub-
discipline of sociobiology. Despite its unattractive connotations,
there are good reasons to retain the earlier term. Evolutionary
psychology currently denotes quite a specific body of thought,
and using the term 'sociobiology' leaves one's options open. On
the one hand, it acknowledges the hardline fitness maximizers,
and the ideological cap that fits them. On the other, it suggests the
possibility of very different inflections, whether kinder and gen-
tler, or of a different colour altogether. The possibilities run the
gamut from Social Darwinism to Darwinian socialism. It is also
franker. What we're dealing with here is not safe, not unprob-
lematic, not easy. But it has far too much potential to be left to the
sociobiologists.

3

The Darwin Seminars ran during the period in which this book was written, from 1995 to the end of 1998, and inform it throughout its course. They were a series of lectures and discussions, held at the London School of Economics, and they were a salon as well. The term has been applied to them both pejoratively and approvingly. One's attitude to salons depends largely upon whether or not one is invited, but these things are an inevitable part of metropolitan life, and science is as entitled to a share of the glamour as arts or letters. The events had another kind of aura, too: they seemed to be bidding – quite consciously – for legendary status, like a nightclub which becomes the hotspot of a new scene; the kind of place people in years to come will pretend they were at.

Organized by Helena Cronin, a philosopher by training, they featured an extraordinary range of speakers. There was Simon Baron-Cohen, who talked about mental modules and autism; David Haig, who discerned in the trials of pregnancy conflicts of genetic interest between mother and foetus; Anders Pape Møller, who glued tail extensions on to swallows to study the effect on their sexual success (positive, at least at first); Garry Runciman, who spoke about armoured infantry in ancient Greece; Robert Trivers, who dedicated his lecture about self-deception to his former graduate student Huey Newton, once a founder of the Black Panther Party; and a lot of scientists who constructed mathematical models instead of doing experiments. The audience was similarly variegated. As well as biologists and psychologists, you might have encountered archaeologists, palaeoanthropologists, the odd novelist, and plenty of local economists.

The latter did not attend simply because they were on the premises. Like the others, they found in modern Darwinism an intellectual lingua franca that allowed people from different disciplines to participate in a common discussion. This seemed like a refreshing change for academe, where the mode of production forces disciplines to become ever more specialized and inaccessible to outsiders. At the same time, the language had a familiar accent. Modern Darwinism, dating from the mid-1960s, is based on the work of a small group of theorists who shook down evolutionary theory and gave it a new set of key themes. The thrust of their efforts was to base Darwinism on individuals, instead of groups; and then to propose mechanisms by which co-operative behaviour could arise from self-interest. Their point of departure came in 1966, with George Williams's *Adaptation and Natural Selection*, which established the principle that nothing in evolution happens because it is good for the species. Williams argued that selection is unlikely to operate at the level of groups, and that the fundamental unit of selection is the gene. Groups are transient and blurred at the edges; each individual created by sexual reproduction is a unique assortment of genes; the only true constant in reproduction is the gene – the 'selfish gene', as Richard Dawkins later called it.

During modern Darwinism's formative period, in the 1960s and 1970s, John Maynard Smith, Robert Axelrod, William

Hamilton and Robert Trivers were prominent among those who worked out how to explain co-operation, between both related and unrelated individuals, without invoking group selection. The notion of inclusive fitness, or kin selection, recognized that individuals shared genetic interests to the degree that they were related. Evolutionary game theory shed light on how unrelated individuals might find ways to practise reciprocal assistance without being cheated. No wonder the economists felt at home in Darwinian circles. Liberal economics and biology were singing from the same hymn-sheet.

At first I was bemused and sceptical. The focus seemed to veer from the grist of molecular biology journals to the stuff of popular psychology magazines. I struggled to grasp the former, and was surprised at the readiness of serious scientists to entertain the latter. A packed hall listened to Devendra Singh, of the University of Texas, give an illustrated lecture about what men like to see in women. Singh's contribution to the search for human universals was evolutionary psychology's equivalent of the Golden Section, a ratio of 0.7 between the measurements of a woman's waist and her hips.

Singh reasoned that it is in the reproductive interests of males to avoid mating with females who are already pregnant. It would therefore be adaptive for them to be attuned to cues indicating whether a female is pregnant or not. Since a woman's waist thickens early in pregnancy, ancestral males who avoided females with unindented figures would have fathered more offspring than males who did not discriminate in this way. Furthermore, Singh argues, a waist wider than the hips indicates susceptibility to illnesses, including diabetes, heart disease and certain cancers. 'Attractiveness, health and fecundity at a glance,' as he put it.[7]

To test for signs that such a process of selection had left its mark on the male psyche, Singh showed men line drawings of women with different waist–hip ratios. His subjects consistently expressed a preference for a ratio of around 0.7, and he found this magic number in all sorts of other places. In Britain, he told the audience, he was always asked 'What about Twiggy?' He had checked the vital statistics of the 1960s model, the forerunner of the 'waifs' of the 1990s, and found that her waist–hip ratio was close to the ideal, at 0.73.

The logic was sound. As the eminent ethologist Robert Hinde commented, Singh had shown that a preference for relatively narrow waists would have served the reproductive interests of ancestral males. But Hinde objected that Singh had not done the experiments necessary to establish his case. I was left with other reservations. Least importantly, I noted with interest that nobody seemed to object to the titillating undertow of the presentation. Across the Atlantic, the science writer John Horgan was also struck by the same lack of reaction when Singh showed his slides at a meeting of the Human Behavior and Evolution Society. 'The headless woman in black leather panties has got to be the last straw,' he gasped, but protest came there none.[8]

There were, however, some doubts expressed about the evolutionary importance of the diseases Singh associated with high waist–hip ratio. These mainly occurred in middle and old age, and in the West, so it seems improbable that they would have had much impact on the number of children women bore in the ancestral environments. They might have had an indirect effect upon the survival of these children, by reducing the amount of maternal care the children enjoyed, but it seems unlikely that they would have been major causes of death among foraging people. Any information that waist–hip ratio provided about their likelihood would therefore have been of questionable reproductive advantage to males, making it less plausible that the mechanisms of selection would have engaged to inscribe a preference for the trait into the male psyche.

Before his talk, Singh had visited the Natural History Museum in South Kensington, to examine replicas of the Palaeolithic figurines known as 'Venuses'. He found that though the figures they depicted were obese, the proportions of waist and hip were in accord with his theory. This seemed to undermine the adaptive value of sexual inclinations based on waist–hip ratio, since it implied that these preferences induced men to overlook obesity, which is unlikely ever to have been a healthy trait. More significantly, though, his discussion raised questions about the interpretation of cultures. First, the evidence is selective. The 'Venuses' are well known; less famous, and less readily readable through the lenses of modern culture or art in historical times, are an equally widely distributed group of female statuettes which are cylindrical rather than globular. Some of them have been found at the same sites as their globular sisters, including Willendorf, where the most celebrated 'Venus' was discovered. The very first prehistoric figurine to be dubbed 'Venus', found at Laugerie-Basse in the Dordogne region of France, has no head, a trunk with almost parallel sides, a pair of legs that give the figure only rudimentary hips, and a mark between them which is the only clear indication of the statuette's sex. A number of the statuettes found in Russia are, literally, stick figures.[9]

Leaving these aside, why on earth should anybody assume that a statue made in the Stone Age has a meaning comparable to that of a pin-up produced under modern capitalism? Just because the dominant use of the female form in our culture is as an object of sexual desire, there is no reason to project our perspectives on to an artefact 20,000 years old. They may have represented matrilineal inheritance; though full-blown matriarchy, proposed by Victorian archaeologists, has now fallen from favour, except among the goddess tendencies in feminism. Some of them may have represented pregnancy, or conditions of plenty in which people might grow fat, or women's power, or something else altogether. According to at least one strand of medical opinion in the earlier twentieth century, the Venus of Willendorf is an anatomical illustration of endocrine obesity, 'an index of the sedentary, overfed life of woman in the prehistoric caves'.[10] One ingenious proposal is that the statuettes represented the artists, women depicting how their pregnant bodies looked as they stood and gazed down at themselves.[11] Another recent suggestion, from the palaeoanthropologist Randall White, is that they were magical objects whose primary purpose was to protect mothers during childbirth.[12] Meanwhile, evolutionary psychology is content to remain on the level of pin-ups.

Assumptions about existing cultures may not be warranted either. To test the possibility that his results merely reflected western tastes, Singh gave his questionnaires to visiting Indonesian students. They may have been more familiar with Western images of women than he supposed, though, particularly since they were Christians rather than Muslims. Under the bombardment of images that would have begun before they got off the plane, they may have rapidly learned to associate a particular kind of figure with glamour and sexual allure; or they might have been inclined to give responses which they understood to be appropriate in their host culture.

This is not to deny that Singh has a hypothesis which can be investigated. But you can't grasp the universals of human nature without recognizing cultural differences. When Singh's drawings were shown to men from a population who have remained more isolated than most, the Matsigenka of southern Peru, higher waist–hip ratios were preferred to lower ones. Among a more Westernized Matsigenka group, men considered lower waist–hip ratios more attractive. And preferences in a third, even more Westernized group from the same region were indistinguishable from those of North Americans.[13]

Anthropologists and archaeologists are concerned above all with context. They become understandably vexed when they hear evolutionary psychologists proposing universal human truths without attempting to set their claims within the context of anthropological or archaeological knowledge. The undercurrent of antagonism is mutual. Evolutionary psychologists are advancing a new paradigm, based on the idea of an evolved human nature, against what they call the 'Standard Social Science Model'. The term was coined by John Tooby and Leda Cosmides, and is sufficiently current to be known by its initials, as though it were a technical abbreviation rather than a rhetorical device. It denotes the view that 'the contents of human minds are primarily (or entirely) free social constructions, and the social sciences are autonomous and disconnected from any evolutionary or psychological foundation'.[14]

Ironically, the mood of much discussion in evolutionary psychology is strongly in favour of keeping the social sciences disconnected from Darwin. 'Standard' social science is regarded as bathwater without a baby, a body of theory with nothing useful to offer evolutionary psychology. The implication is that adequate new social sciences can be built entirely within a Darwinian paradigm. Many evolutionary psychologists will naturally protest that this is a caricature of their position, or deny that such an attitude exists. But although one commonly encounters derogatory references to the SSSM in evolutionary psychological discussions, it is much rarer to hear calls for partnership between evolutionary thinking and established social science, or for integration between them. The assumption seems to be that there is only room for one paradigm.

A good example of how pluralism can work in practice comes from the work of Martin Daly and Margo Wilson, whose book *Homicide* is often, and justifiably, cited as a showpiece of evolutionary psychology.[15] The Darwinian structure of their thought is illustrated most clearly by their work on the killing of children by family members. They chose killing, not because of its drama but because of the relative reliability of statistics on it. Kindnesses within households are harder to measure than abuses, and the harm least likely to go unreported is homicide. Daly and Wilson reasoned that since step-parents did not share a lineage of genes with their step-children, they would be more likely to harm them

than would biological parents. They found their hypothesis spectacularly upheld. Their findings, which draw upon figures from
Canada, the United States, England and Wales, show a consistent
pattern. For babies and very young children, the risk of being
killed by a step-parent is between fifty and a hundred times
higher than that of death at the hands of a genetic parent. They
have concluded that being a step-child is 'the single most important risk factor for severe child maltreatment yet discovered.'[16]

One possible objection to their analysis is that this reflects
upon the kind of person who becomes a step-parent, rather than
upon the operation of mechanisms favouring inclusive fitness.
But when violent step-parents live with both step-children and
their genetic children, they tend to spare their own kin. In any
case, the finding stands as an empirical indicator of risk, which can
be taken into account when social services are considering a
child's welfare. If a child is at increased risk through living with a
step-parent, it is important to know this, regardless of the underlying reasons for their vulnerability.

This point also addresses another criticism, often levelled at
evolutionary psychologists: that they are merely stating the obvious. Suspicions of this kind had nagged at me through the Darwin
Seminars until I heard David Haig make sense of phenomena in
pregnancy – high blood pressure, diabetes, massive washes of
hormones – that are physiological, marked, and hard to explain
other than as struggles over resources between foetus and mother,
arising from conflicts of genetic interest. It ran completely against
intuitions based on conventional assumptions about the intimate
bond between a mother and the fruit of her womb, yet I found it
immediately convincing. Others did not. In a letter to the *New
York Times*, Abby Lippman called it an 'outrageous' suggestion
which 'betrays the astonishing assumption of patriarchal societies
that fetuses are separate entities that happen to grow inside
women's bodies'. Lippman, a professor of epidemiology, insisted
that the uterus is not 'the site of a protracted battle between

competing genomes, but the nourishing, life-sustaining part of a woman's body, where mother and wanted fetus cooperatively coexist.'[17] So anxious that it felt the need to assert that every foetus is a wanted foetus, her rhetoric overwhelmed its own efforts to mount what could have been a salutary critique of scientific metaphor, and failed to provide alternative explanations for the phenomena Haig had adduced as evidence.

Daly and Wilson's response to the charge of truism is that if it was so obvious, why did the social sciences fail to notice it? Daly and Wilson sound like Darwinian partisans when they talk about the failure of standard social science to realize that folklore and folk wisdom were right on this score. When they began their investigations in the 1970s, they tartly remarked in their essay *The Truth about Cinderella*, 'most of those who had written on stepfamily conflicts apparently believed that the problems are primarily created by obstreperous adolescents rejecting their custodial parents' new mates'.[18] Nobody had asked whether having a step-parent increased a child's risk of being injured. 'It's got to be more than merely an oversight, it's got to be motivated neglect,' Daly declared in an interview.[19]

Some of their other remarks might also have caused uneasy twitching in liberal circles. They had observed that US inner cities show an anomalous pattern of homicide rates among couples, compared to the suburbs or cities in their native Canada. Typically, men kill their female partners several times more often than women kill their men. In this dimension of US inner city death, however, sexual equality had been achieved. Daly and Wilson first thought that the prevalence of guns might be the reason, but they found that women tended not to use firearms to kill their partners, and killings by other means show the same sexual balance. The pattern was not the same across all ethnic groups: Latina women almost never killed their husbands.

Daly and Wilson did not, however, explain this difference in terms of racial character. They saw it as the product of social

relations. Latino communities are, in anthropological jargon, patri-
lineal and patrilocal. This means that material resources are
transmitted through the male line, and men stay within their communities instead of leaving to join others when they enter
marriage or similar relationships. Latina women are thus surrounded by their partners' relatives and isolated from their own.
Among urban African-Americans, by contrast, the central resource
of housing is transmitted down the female line, as women get
entitlements to leases via their mothers or sisters. Black women
gain strength from being in networks of kin, whereas black men
lack equivalent support. There is also a higher turnover of partners among African-Americans than among the Catholic Latinos,
and so more step-children. Some of the domestic homicides may
have occurred when women attempted to defend their children
from attack by step-fathers, Wilson speculated, observing that
there were no shelters available to these women, nor did they
have effective access to help from the police. Killing their partners
could be seen as a form of 'self-help'.

Daly and Wilson started out studying the behaviour of rodents
and monkeys, and still maintain a research interest in the rodents.
'Perhaps we're pre-adapted to treat *Homo sapiens* as just another
critter,' Daly observed. That's the kind of talk that makes anthropologists shudder – particularly when the *Homo sapiens* in
question are black humans at the lower end of the social scale,
being discussed by white ones comfortably lodged in the upper
reaches. Anthropology and sociology are only too painfully aware
of the oppressive potential inherent in such analyses. Frederick
Goodwin had to resign his post as chief of the US Alcohol, Drug
Abuse and Mental Health Administration in 1992, after commenting that it might not just be 'a careless use of words' to refer
to inner cities, which had lost some 'civilizing' elements, as 'jun-
gles'.[20] For those who take the sociologist Durkheim rather than
Darwin as their intellectual patriarch, the savannah may seem
just as dubious as the jungle.

Given the state of theoretical play in the social sciences at the moment, it is easier to talk about representations than solutions. Talking to Martin Daly and Margo Wilson, I became confident that they were instinctive liberals, in disposition as well as ideology – and in contrast to the brittle liberal veneer that some other sociobiologists feel obliged to adopt. It wasn't just that they referred to the fact that Canada has nationalized healthcare, which mitigates inequality and leaves fewer of the poor in a state of desperation, as a factor relevant to the pattern of domestic homicide. It was that they seemed to be genuinely pluralistic thinkers, who wanted to get Darwinism working productively together with other schools of knowledge, rather than trying to replace the social sciences with biology. People frequently imagine Daly and Wilson are proposing that the murder of step-children is an adaptive act, in which individuals dispose of rivals for their own children, actual or potential. Such things happen among langur monkeys and lions, but although Daly and Wilson recognize that humans are also animals, they do not make the mistake of thinking that we are the same as particular animals. Nor do they draw a conservative moral from their statistics, to argue that step-families are a pathological departure from monogamy. They point out that step-families were common in the last century, and probably throughout human history, because people remarried after the early death of spouses. Ancestral environments, they believe, may have encouraged a readiness to take on step-families. Being a parent is extremely difficult: when the child is one's own, evolved mechanisms underpin the commitment that is necessary. Without the support that comes with kinship, it is much more likely that an adult will fail to meet the commitment. The homicide statistics reflect the most extreme instances of this failure.

Another way of putting it is to say that step-parents are less likely to live up to their roles. Daly and Wilson argue that this is an insidious metaphor which encourages us to think that life is theatre. They also consider it to have unwarranted intellectual pretensions. 'There is no theory,' said Martin Daly, '. . . It seems to be nothing but a metaphor.' Yet they both acknowledge that talk of roles has produced the notion of 'scripts', which they accept can be a useful tool for making sense of how families work. It seems to be the utility of the idea that counts, not where it came from. Margo Wilson summed up their discussion of spouse-killing in US cities by pointing out three principal factors: matrilineality, step-families and poverty. The first is a concept derived from anthropology and primatology; the second comes from their reading of evolutionary theory; the third is a sociological and political perspective. They make sense together, and they lead to an analysis which the individual currents could not reach on their own.

'We've got to synthesize evolutionary understandings and cultural understandings, not pit them against each other,' said Daly. Daly and Wilson's theory, which has led to studies in women's refuges and new tools for social work, illustrates what such syntheses can achieve in practice. You can start from rodents and still end up on the side of the angels after all.

4

My fascination with prehistoric humankind was kindled a few years ago in a visit to a small group of caves at Cougnac, in the French department of Lot. The drawings on the cave walls were thought to have been made around 18,000 or 16,000 years ago, the better part of 10,000 years before people settled down to farm the land. It seemed an extraordinary span, though in terms of human origins it rapidly comes to seem like the day before yesterday (as do the revised dates, ranging as high as 25,000 years, which have subsequently been obtained from tests performed on the pigments).[21]

In one of the chambers I was able to spend a few moments on my own, standing an arm's length away from an image of an ibex, a relief created from ochre and the natural contours of the rock. In that arm's length was the dizzying thrill of simultaneous presence and unimaginable distance. To stand on the brink of such an image, on the verge of communication with a mind more than 500 generations away, was as enthralling as the sight of the stars in the heavens.

I had put myself in the artist's place, standing where the artist must have stood. I could imagine reaching out, in the light of a guttering oil lamp, and making a line on the rock. Within this narrow frame of vision, I knew roughly how it had looked to the artist. Beyond that, it seemed, all else was a shimmering curtain of maybes. Maybe the artist was a man, maybe a woman; maybe the ibex was part of a ritual, maybe an image of beauty, maybe a record of an event, maybe something beyond modern imagining.

To put oneself in another's place is a fundamental operation of consciousness, requiring theory of mind. Primatologists debate whether apes have it; developmental psychologists wonder when children acquire it. This book explores theories of mind in a different sense; scientific theories about ancient minds. The impulse behind it, though, is the fascination of the possibility that we can achieve theory of mind across geological time; that we can put ourselves, to however limited a degree, in the place of our ancestors.

At Cougnac, new to prehistory, I thought that all one could do was make up stories. Since then, I have learned that although certainty about human origins is an endlessly receding horizon, there are devices we can use to take us closer to that horizon.

5

Astory need not be just a story. A theory need not be just a theory. Formally, evolution may be a theory, but it has consistently passed the tests set it, and to all intents and purposes it is the plain truth. The palaeontologist Stephen Jay Gould is clear on this point, but he is profoundly sceptical about the application of evolutionary theory to human behaviour. He has endorsed the project in principle: 'Humans are animals and the mind evolved; therefore, all curious people must support the quest for an evolutionary psychology' – but not, it appears from the rest of his article, if the quest is being pursued by evolutionary psychologists.[22] Following his cue, many among the lay public have learned to presume that evolutionary stories about people are 'just-so stories', a dismissive phrase he has popularized. Sometimes, when scientists are in a self-deprecating mood, they also refer to 'the stories we tell', as if they were just stories.

Gould has also drawn his readers' attention to the importance of checking original texts, instead of relying upon secondary accounts of them. Rudyard Kipling's original *Just-So Stories for*

Little Children appeared at the turn of the century, at the zenith of the British Empire and the nadir of Darwinism, when many scientists had lost interest in Darwin's idea. Since then there has been a reversal of fortune, dating in fact from around that time, when scientists rediscovered Mendel's work on the mechanism of genetic inheritance. Darwin is now held in awe, if not in universal affection, while Kipling is widely regarded as a deplorable old imperialist, although his poem 'If' remains a favourite with the British public. Not all his lines remain pleasing to the modern ear, though. 'The Ethiopian was really a negro, and so his name was Sambo,' he jocularly confides to the little children in 'How The Leopard Got His Spots', having previously used a worse word than 'negro'.[23]

Many people would see remarks like these as pollution, which foul the rest of the text like spit in a glass of milk. The only safe and proper reaction is to throw the lot down the drain. As the century has begun to wind itself down, though, a new perspective has arisen in certain quarters. Reconciliation and mutual understanding have come back into favour. In that spirit, perhaps it might be instructive to apply a different model to unsatisfactory writings of the past. Instead of seeing a text as a fluid, in which any toxic trace contaminates the whole, it could be considered as an assembly of components. Software is the obvious analogy, which implies that it can be debugged. If the flaws are really extensive, then one constructive option might be to strip out the serviceable elements and incorporate them into a new architecture. Maybe the *Just-So Stories* could be turned into something more than just stories.

Not all of them are suitable for modernization. Some are to be taken as one finds them, since one either accepts the morals they are designed to illustrate, or one doesn't. The rhinoceros stole the Parsee's cake, so the Parsee rubbed cake crumbs into the rhino's hide, and that is how the rhinoceros got his rough and lumpy skin. The camel said 'Humph!' every time he was urged to work for men, as is the duty of all animals. It seems a rather restrained response, especially for a camel, but a djinn punished him by giving him the hump.

'How The Leopard Got His Spots' is a somewhat different case, being more concerned with telling a story than pushing a moral, and it contains a grain of truth in the wrapping of a fable. It begins in the High Veldt of South Africa, a desert habitat dotted with tufts of grass. Like all the other features of the environment, the grass is a uniform 'yellowish-greyish-brownish colour', and so are the animals. The herbivores include giraffe, zebra, eland, kudu, bushbuck, hartebeest and the now-extinct quagga. Their predators are leopard and human, the text being taken from Jeremiah chapter 13, verse 23: 'Can the Ethiopian change his skin, or the leopard his spots?'

In those days, animals lived a very long time, and eventually the prey learned to avoid the predators. Some of the herbivores migrated to a forested environment, in which the foliage produced an uneven illumination pattern of 'stripy, speckly, patchy, blatchy shadows'. To avoid standing out in these conditions, the animals developed camouflage patterns; the blotches of the giraffe, the zebra's stripes, delicate lines like bark on the eland and kudu. 'Can you tell me the present habitat of the aboriginal Fauna?' the Ethiopian asked Bariaan the Baboon, archly acknowledging the natural sciences. Bariaan, a regional oracle, replied that the aboriginal Fauna had joined the aboriginal Flora. Eventually, the predators worked out from Bariaan's Delphic remarks that they needed to move into the forest, and adopt similar cryptic colouring.

Kipling invokes two agencies by which external appearance can be changed. One is Lamarckian, whereby characteristics acquired during an organism's lifetime are transmitted to future generations. The other is magic, whereby an organism can change its skin at will. Natural selection it isn't, but adaptation it is.

If we are to be positive here, and try to build on this shaky point of contact with science, we must first drop Lamarck and magic. In the process, we can modify the awkward device of longevity employed by Kipling. Instead of individual animals living a very long time, we need a breeding population to persist long enough for new traits to arise, through the effect of selective pressures acting upon variations that appear within the population. Herbivores that are blessed by genetic chance with blotches or stripes will be less vulnerable to predators; they will thus leave more offspring than their unpatterned companions; over the generations, the traits will be developed and fixed. Similar pressures produce spotted leopards.

Next, the ecology needs to be examined. A gaping hole appears immediately. If the herbivores were happy in the desert, presumably subsisting on a diet of yellowish grass, how did they adapt to the very different conditions of the forest? Evolutionists sometimes refer to pre-adaptation, in which a trait that has evolved for one function turns out to be useful for something else when circumstances change. Stretching a point, you could argue that the giraffe's neck was pre-adapted to the forest habitat, but that begs the question of what the beast was doing on the treeless veldt in the first place. Even if the other species also proved adaptable to wooded conditions, the radical change of environment would surely have a big impact on their way of life. Some of the most important insights of behavioural ecology concern the effect on social relationships of how resources are distributed; whether dotted in patches across the landscape, as in the desert, or evenly spread, as in a lush forest.

At the same time, reality checks are needed. Depending on what sources are available, these may include reference to the fossil record, and to whatever information is available about environments in bygone times. They will also include comparisons with what is known about the behaviour and distribution of living members of the species in question.

Finally, it's desirable to run an audit for each adaptation, weighing the benefits to the individual against the costs. Hanging in the balance are the costs of finding food in each environment, the likelihood of falling victim to a predator, the amount of competition for resources in the different environments, and the relative concentration of parasites. In many respects the forest may be more congenial. There will be more food, in greater density. The climate may be less extreme, making it easier for an animal to maintain an adequate intake of water. There will be less need to defend against carnivores by gathering together in large groups, one of the main tactics employed in open country. This may reduce conflicts between members of the same species. On the debit side, forests tend to contain more creatures from a wider range of species than other habitats. Many of these are parasites of various unpleasant sorts, from ticks to viruses. And although food may be more plentiful, there will be more competitors for it, employing a wider range of tactics. For the animals in Kipling's scenario, it would be like moving from a small village to a big city.

Children a century on from Kipling do not just read stories. They play simulation games, using computers, and that is what scientists do to examine evolutionary or ecological issues. It would be a relatively simple matter to assign some values to the different factors in the Kipling scenario, and use them to animate a little world on a computer screen in which animal icons move from veldt to forest. With the microprocessors calculating the costs and benefits, the rates of change under various conditions could be observed. If the system reached an equilibrium containing stable populations of all the species, then Kipling's ending, in which everybody lived happily ever after, could be said to be plausible.

It would also be possible to specify the range of conditions under which the Kipling scenario could work. Altering the simulated climate by changing the temperature settings, for example, would have an effect on the density of the vegetation. At some point the forest would cease to be able to support the herbivores. This level could then be compared with palaeoclimatological data,

74018

to give an idea of when the leopard might have got his spots, as well as how.

By this point, the tale of how the leopard got his spots would have become more than just a story, it would now be a working model. A model is more accountable than a story. It is governed not by a despotic narrator, but by the rule of law. The elements within it are not controlled directly by the narrator, but act according to the rules of the game. If a storyteller chooses to disregard a character, that character ceases to play any part in the story; whereas the elements of a model carry on interacting whether the model-builder is paying attention to them or not. A model can be set rules which are shared by other bodies of knowledge, allowing it to be connected to them and rendering it transparent to external scrutiny. Stories can be told; models can be tested. That's science; and that's the tale of how one story gets to be better than another.

Some people might object that processing texts in this way takes the poetry out of them. The loss might not be too great in the case of Kipling, who 'wrote poetry like a drill sergeant', according to W. H. Auden. It is true that the human touches are discarded as the narrative is stripped down to its structural elements, and that science aspires to be impersonal – officially, anyway. But poetic structures are not the only ones capable of containing the ions and fluxes of charged imagination. Although the sketch for the Leopard's Spots model employs pretty conventional ideas about camouflage, one of the species in the model has inspired scientists to bolder flights of reason. Nearly all zebras live in substantial herds on arid plains, where vegetation is sparse. There is nothing for their stripes to blend in with, except the rest of the herd. One school of thought holds that the effect of the stripes is to make it hard for an attacker to distinguish one zebra from another, and therefore to choose an individual as a target. An alternative interpretation comes from Amotz and Avishag Zahavi, a particularly bold pair of evolutionary thinkers. 'If the stripes were meant purely for camouflage, they could have been random, like a leopard's spots,' they argue. The stripes act not to draw attention away from an individual, but to accentuate parts of that individual, including lips, hooves, legs, neck and rump. These markings may send messages about the animal's condition to a variety of receivers, both friendly and hostile. 'A predator looking for easy prey, a rival evaluating his chances in a contest, a female looking for the best father for her offspring, all can benefit from evaluating the muscles of the rump.'[24]

The Zahavis' remarks about the zebra occur in a spectacular account of signal evolution, of which much more later. For now, it's an opportunity to note that their 'Handicap Principle' is a clear example of how a story can be turned into a model. When Amotz Zahavi first published his ideas, he was more successful in piquing interest than winning acceptance, which did not come until other theorists published mathematical demonstrations that the Handicap Principle might work.

The critical stage in building a model is not applying the algebra. An idea does not have to be translated into numbers, though if that can be done, so much the better. What really matters in assessing a proposition about evolution is that both the benefits and the costs are taken into account. A story that weighs this balance will have a tautness missing from one which simply suggests an adaptation that it might be nice for an organism to have. A just-so story is like the M25, London's orbital motorway. On a true motorway, lane procedures are the keys which keep the system taut and sprung. Each driver is supposed to drive along in the lane nearest the verge, moving into the next lane to overtake, and then into the lane nearest the central reservation if necessary. When the traffic is not too heavy, the aggregate effect of drivers following this simple procedure is to maximize the efficiency of the system, allowing most vehicles to travel as near to the speed limit as they want to go. It gives motorway driving a shape, and it gives the driver a sense of purpose. As the weight of traffic increases, its effectiveness diminishes and the logic of motorway driving is undermined. On the M25, congestion is so bad that the collapse of the system has been institutionalized. Drivers are instructed to stay in the same lane, removing the *raison d'être* of a motorway. The M25 is just columns of traffic running next to each other. A road in which the vehicles do not really interact is like a just-so story whose agents do not really interact.

Another way in which many just-so stories resemble the M25 is that they are circular. A good evolutionary story has the sprung tension of a clear highway, and it goes somewhere.

The Prime Maxim of evolutionary thinking:

Always consider the costs.

No, it is not coincidental that cost–benefit Darwinism is flourishing in a period during which fiscal rectitude has been sanctified as the supreme public virtue, by governments with a deep aversion to levying taxes. This new moral order is affirmed everywhere, from the office manager's memos to the dictates of the World Bank. In the public sphere, it affirms the individual and devalues the collective, just as neo-Darwinism affirms that selection acts on individuals, but not on groups.

There are still pockets of resistance. Quoted in John Brockman's collection of interviews with scientists, *The Third Culture: Beyond the Scientific Revolution*, Lynn Margulis rejects not just the modern Darwinism of the last thirty years, but also the older 'neo-Darwinism' which began to emerge in the 1920s. According to accepted wisdom, the link that Darwin missed was the connection between his theory and Mendel's model of genetic inheritance. One of the great missed opportunities in the history of ideas is the copy of Mendel's paper still preserved in Darwin's library, its pages uncut. Neo-Darwinism made good that omission.

According to Margulis, though, the two theories are incompatible. She compares neo-Darwinism to phrenology, and predicts that it 'will look ridiculous in retrospect, because it is ridiculous'.[25] She recounts how she asked the biologist Richard Lewontin why 'he was so wedded to presenting a cost–benefit explanation derived from phony human social-economic "theory"'.

Margulis's own story is, in the apt expression Daniel Dennett uses in the comment which follows her interview, delicious. She had a brilliant idea which anybody could grasp, but which the scientific establishment laughed at. Now her peers have accepted that she was right. Up to a point, that is. She is still living in a paradigm of her own. Her grand theme is that change in living things happens by merger rather than by mutation. One of the most profound transformations in the evolution of life was the appearance of cells with nuclei. This may have occurred, she argues, through a predatory invasion of one bacterium by another. Assault turned to alliance, as the two forms integrated themselves together. Nucleated cells also have structures called mitochondria, which serve as a source of energy. Though they function as sub-cellular organs, or 'organelles', mitochondria have their own DNA and look like cells within cells; which is exactly what Margulis says they are.

On mitochondria and nuclei, Margulis's ideas are now accepted wisdom. Her peers remain unpersuaded that such processes are the dominant mode of evolutionary change, or that mutation cannot provide enough material upon which selection can act. In their quoted comments, Daniel Dennett and George Williams both make counter-charges of ideological bias. Dennett expresses regret that, in his opinion, Margulis is trying to politicize the idea of symbiosis, as a tool to promote co-operation instead of competition. Williams points to her support for James Lovelock's Gaia hypothesis, which proposes that life on Earth behaves as a united entity to maintain the environmental conditions that support it. He suggests that 'she wants to look out there at nature and see something benign and benevolent . . . Whereas I look out there with Tennyson and see things red in tooth and claw,' he remarks.

'Time will tell,' Williams concludes, 'and will show that my approach is more fruitful in generating predictions about discoveries we're going to make.' In a scientific exchange of views, that is the last word to have. Predictive power is the measure of a theory or a research programme. It is a way of saying that an idea can be put to good use.

This sense of practicality is one of modern Darwinism's most attractive features. Stephen Jay Gould calls adaptationism 'the British hang-up', and sees Britain as a hotbed of 'Darwinian fundamentalism'. British theorists tend to operate on the presumption that any trait of an organism is an adaptation produced by evolutionary selection. On the continent of Europe, theoretical biology has traditionally paid more attention to the structure of organisms. Somebody like Brian Goodwin, the Open University biologist who argues that the influence of structural logic on living forms has been massively underrated, is a continental thinker who finds himself on an uncongenial island. Gould's sympathies are with the continentals.[26]

Maybe the heterodox evolutionists are in possession of some vital truths. Perhaps the broad vision of Lynn Margulis will eventually be accepted, as well as the nuclei and mitochondria. Possibly the evolutionary theorists David Sloan Wilson and Elliott Sober are right to argue that group selection really does happen after all; or possibly John Maynard Smith is correct to say that the differences between them and orthodoxy are only semantic.[27] But one distinguishing characteristic of orthodox, reductionist neo-Darwinism is its productivity.

It even has something to say about Gaia. According to James Lovelock and his colleagues, algae in the sea produce clouds. Directed by the Gaia hypothesis to look for a compound which might transport sulphur, an element essential to life, scientists have discovered that marine algae emit a gas called dimethyl sulphide. Reaching the air, this could react with oxygen to form tiny solid sulphate particles. Water vapour could condense around these seeds to form clouds, which would reflect sunlight away from the surface of the planet, and thereby help to cool it. Warmer temperatures would cause the algae to flourish, producing more dimethyl sulphide, which would stimulate cloud formation, and lower the temperature again. Algae thus act as a living thermostat; cooling the Earth by as much as four degrees, according to one estimate.[28]

The difficulty for conventional evolutionary theory is that it's easy to see how algal dimethyl sulphide production might be good for the planet, but less easy to see what is in it for the algae. William Hamilton, one of modern Darwinism's most influential theorists, became intrigued by the problem. A chemical precursor of dimethyl sulphide had a possible use as an antifreeze – but that would not explain why it is produced by algae living in warm seas. Hamilton speculated that the cells might need protection if they were to be transported high into the air. If algal cells could get airborne, they could take advantage of the winds to disperse, like pollen. Together with Tim Lenton, an atmospheric chemist, Hamilton worked out a hypothesis in which the crucial role of dimethyl sulphide is to create upward air currents. This it does by producing the sulphate particles; as water vapour condenses around them, heat is released, warming the air, which duly rises. Algal cells already airborne are carried with it, and the breezes that the air currents create at sea level help draw algae from the surface of the sea into the air. Many algal blooms are known to be clonal, each cell containing the same genes, so the genetic interests of a cell up in the clouds might well be identical to those of one in the waves below. The model thus illustrated the possibility that what might be good for one algal cell might be good for many others, and for the planet as well. Gaia and modern Darwinism might be reconciled – though it's not exactly what Lynn Margulis had in mind when she spoke of mergers and symbiosis.

In their very different ways, the cases of the algae and the step-children powerfully suggest that modern Darwinism can be integrated with other ways of understanding the world. They also show that the formal requirements of evolutionary theory can be what imagination needs to support its flights, rather than constraints which keep it earthbound. Nor is it necessary to accept the rules as final truth. You can accept that reductionism has its limits, while sticking faithfully to reductionist principles when working with scientific ideas. Reductionism gets results – many more, so far, than rival methods – and these results are woven together by the common understandings created by the rules of the programme.

Humans are always close to their biology. They are of course inseparable from it, always animals as well as people, always living through both their nature and their nurture. These pairs are not opposites, just different aspects of the same thing. Perhaps, at the turn of the new century, this may be a good time to turn over a new leaf; to go beyond the opposition between soul and flesh that is so fundamental to Christian tradition. Descartes's dualism of mind and body usually appears in histories of science as a superseded concept: perhaps now is the time to blow away its lingering spirit. When both aspects seem important in addressing a question about human affairs, it should be possible to take them as a whole. It will not always be a harmonious whole, but its tensions should be creative.

Sex is the natural theatre in which to work some of these tensions through. As well as being a matter of public interest, it is what makes the world of evolutionary psychology go round. The idea that the sexes have different reproductive interests is profoundly simple, like Darwin's original idea of natural selection. Unlike Darwin's world-transforming idea, differences of reproductive interest have been recognized from time immemorial. Modern Darwinism tapped the potential of the insight by placing it in a scientific matrix. Its ability to sit happily within both folk and scientific wisdom has since given evolutionary arguments about sex ready access to the media at large. Inevitably, one of the results has been a spate of answers to questions such as 'why men don't iron' (the title of a British television series on evolutionary psychology) which assure the public that science is now back in tune with folk wisdom about human nature.

Folk wisdom is based on generalization, not variation. If most men don't iron, folk wisdom decrees that men don't iron, period. A man who does iron is therefore anomalous at best, and at worst is not considered to be a proper man at all. Instead of embracing a spectrum of behaviours within the normal range, accounts in the folk idiom depict a stereotypical norm and pathological deviations from it. *Why Men Don't Iron* presented a woman who enjoyed building model railways – and had an abnormally masculine hormonal constitution. Although the dominant cultures in North America and a number of European countries take a relaxed view of men who iron, the normative influence of folk wisdom persists even in these regions. Stridently conservative accounts like *Why Men Don't Iron*, relentlessly declaring that efforts to challenge gender roles have simply highlighted the irreducible differences between male and female minds, are accepted with little demur.

Evolutionary psychology is capable of a broader vision. Understanding variation as it does, it can appreciate that behaviour will cover a spectrum. Some men will do all their own ironing, some will do none, but all of them fall within the normal range. In its more elaborate forms, based on game theory and John Maynard Smith's concept of the evolutionarily stable strategy, evolutionary psychology may take the view that it is a good idea for males to be able to pursue different strategies. If all of them are competing to be Iron Men, hard as nails, then the hardest will take all and the rest will be disappointed in reproductive success. It would make sense for some to compete as Ironing Men, offering resources other than tough genes and protection similar to the kind given by the Mafia.

Having set itself the task of describing a universal human nature, though, evolutionary psychology gravitates towards the norm. Peering at the tumult of human behaviour, its eyes light upon what makes evolutionary sense. That leaves a vast amount, much of it what many people would probably consider most interesting. An easy way to make an inventory covering a significant proportion of these all too human foibles would be to consult a few contemporary art catalogues, or similar postmodernist texts. These speak of gender rather than sex, homosexuality rather than heterosexuality, polymorphous perversity rather than reproductive success; of cyborgs made of flesh and silicon, rather than humans adapted by natural selection; of what women could be, rather than what evolutionary psychology says they are inclined to be.

In its own way, through its contortions, obfuscations and prevarications, postmodernism is trying to keep the flame of possibility alive. That is why it is confined to the world of academic discourse, while the world is governed by ideologies of limits. Among the wealthy democracies, where the proportion of voters who pay taxes is high, public spending is strapped by the ability of political parties to offer lower taxes than their rivals. The leadership of Britain's Labour Party has this insight carved on its collective heart. One of the intellectuals closest to it, Geoff Mulgan, has long had an interest in evolutionary theory. 'The ideas contained within liberalism, the Enlightenment and Marxism that humans can make themselves what they want to be – an argument that reaches its apotheosis in Foucault with the image of humans as pure self-creation – is extremely dangerous,' he told Kenan Malik in the *New Statesman*, six months before Labour won the 1997 general election and he was appointed to the Downing Street Policy Unit. '. . . Just as ecological understanding has shown there are all sorts of external limits to what humans can do, so evolutionary psychology shows parallel internal limits, which we transgress at a high cost.' Achieving androgyny would be expensive, for example, since it would be difficult to make men less promiscuous or territorial.[29]

At the time he made these remarks, Mulgan was director of the Demos think-tank. Demos devoted an issue of its journal to 'the world view from evolutionary psychology', demonstrating among other things that the popular media do not have a monopoly on glibness. The preamble to one article announced that according to evolutionists, politicians 'are simply striving to increase their sexual capital'.[30] It took its cue from the text, in which the evolutionary psychologist Geoffrey Miller asserted that 'People respond to policy ideas first as big-brained, idea infested, hypersexual primates, and only secondly as concerned citizens in a modern polity'. An evosceptic insists that what is uniquely human comes first; an advocate of evolutionary psychology says we are primates first and human second. And yet it's not a race, or a zero-sum game in which the social sciences' loss is Darwinism's gain.

In its eagerness to do its job and stop the reader turning the page, the preamble to Miller's piece blurted out what some evolutionary theorists seem to feel but won't quite say; that in a fundamental sense human affairs really are simple, and that sociobiological explanations for them will suffice. Coupled with a disdain for the Standard Social Sciences Model, this adds up to another claim of priority. Resolving this contest will take more than homilies about the virtues of co-operation and dialogue, since the differences are profound and the debates have a long way to run.

One way forward might involve thinking less categorically about questions and answers. It should be possible to agree that science could form part of a perspective which has more than two dimensions.

6

If there is one thing the study of human evolution has in over-whelming abundance, it is uncertainty. It buzzes with agitation; it is forced, whether it likes it or not, into constantly revolutionizing itself. The exhilarating thing about human origins is that they change so fast.

During the period over which this book was written, at least half a dozen landmark discoveries have been announced, though some may not stand the test of time. From Ethiopia came the news of stone tools two and half million years old, archaeology's oldest specimens to date. From a German coal-mine, archaeologists unearthed three spears, calculated to be 400,000 years old. No complete spears of such age had been found before, and their tapered design was a striking demonstration of the intelligence that hominids were then capable of applying to their tools. From Slovenia came a perforated bone, with an estimated age range of 43,000 to 82,000 years, that its discoverers argued might have been a flute. If it really was a flute, not just a bone bitten by an animal, then Neanderthals must have been musicians.[31] Last and far from least, an australopithecine skeleton was discovered at Sterkfontein, in South Africa. It was older and more nearly complete than 'Lucy', the famous specimen found in Ethiopia in 1973.[32]

Many of the islands of the Indonesian archipelago have at some periods been part of a continuous south-east Asian land mass. Others, like the Wallacean islands east of Bali, have been separated from the mainland by deep water throughout the period of hominid existence. In 1998, the journal *Nature* published a paper giving dates of 900,000 years for a collection of stone tools found on one of the Wallacean islands, Flores.[33] Up to this point, the oldest evidence that hominids had crossed long stretches of water was the settlement of Australia, estimated to have taken place between 40,000 and 60,000 years ago. If accurate, the dates imply that the south-east Asian hominids had far more sophisticated technological and planning abilities than their simple stone tools suggest.

The fossils have sprung surprises as well. In the 1994 edition of the *Cambridge Encyclopedia of Human Evolution*, Chris Stringer suggested that a collection of remains found at Ngandong, in Java, were 'perhaps as young as 100,000 years old'.[34] Two years later, new test results were reported. These indicated that the fossils were half that age at most, and possibly as little as 27,000 years old.[35] At the other end of the Old World, dates for the last Neanderthals have been edging forward from 35,000 years. Taken together, the revisions suggest that as late as 30,000 years ago, there may have been three species of the genus *Homo* in existence. *Homo sapiens* was established across the Old World, but Neanderthals persisted on the western fringe and *erectus* on the eastern. Humans as we know them may not have been alone for as long as we had supposed.

In Spain, researchers working on some of the oldest European hominid fossils proclaimed them to represent a hitherto unknown species. That was routine, however, compared with the dramatic announcement that a stretch of DNA had been isolated from a Neanderthal skeleton. Not just any Neanderthal skeleton, either. This was the actual one from the cave in the Neander Tal, or valley, discovered in 1856. Although it was not the first Neanderthal fossil to be found, it was the first to be recognized as a representative of an extinct form of human, and became the 'type specimen' upon which the Neanderthal category was based. The stuff of genetic inheritance is not durable, and whatever they said in *Jurassic Park* – or in the scientific papers that came in its wake, with unsustainable claims to have extracted DNA from insects trapped in amber – it is unlikely that DNA can be recovered from fossils more than 100,000 years old. While many scientists reacted to the single sample (taken from the upper arm) with caution at best, the fact that identical results were obtained in two different laboratories indicated that the DNA came from the Neanderthal rather than a laboratory worker. Whether it was degraded or not is another question, but the sequence supports the view that modern humans are not descended from Neanderthals. The differences are too great.[36]

The Neander skeleton has thus served as the foundation stone of two eras in the study of human origins. It was the example that established the reality of ancient beings who were human, but not as we know it. Now it has inaugurated a new collection of ancient hominid material, comprising strands of DNA rather than fragments of bone. Unfortunately this is likely to remain a small collection, since sufficiently recent fossils often come from tropical regions, where DNA is very unlikely to survive. Even if the Ngandong remains are only 30,000 years old, it would be too much to hope that they can tell any genetic tales about *Homo erectus*. All the same, the human family now belongs to the exclusive club of lineages which can lay claim to genetic information regarding extinct members, along with elephants, who qualify thanks to the mammoths sealed in the Siberian permafrost for up to 100,000 years, and sloths, some giant specimens of which have been preserved in cold South American climes for 10,000 or 20,000 years.

This has a significance deeper than the details of the short DNA sequence extracted from the Neanderthal's arm. DNA is not the repository of ultimate truth that its popular image suggests. But information about it is a fundamental element of what we can know about a form of life. Now that we have a scrap of this to go with the knowledge that comes from Neanderthal fossils, artefacts and the sites where they have been found, the smallest of windows has opened at the molecular level of understanding. Its importance may be symbolic, but that symbolism applies to a form of life that was profoundly like and profoundly unlike us.

The eyes of the world tend to turn to the heavens in search of such beings. We do not know if there is intelligent life out there in the stars. There probably is, but that is not the same as believing 'we are not alone'. The universe is immensely large. This implies that there may be many homes for thinking beings across the universe, but they are likely to be spread vanishingly thin. So the chance that intelligent aliens have visited our planet (even assuming they would want to) is as astronomically improbable as the existence of such life forms is astronomically probable. For all we will ever know, we are alone . . . but the fossil record proves that this was not always so. While Neanderthals and other hominids may not have demonstrated all the capacities that we now associate with humanity, they were undoubtedly creatures with intelligences of their own. Separated from us in time rather than space, they are our own true aliens.

TWO

Symmetry

1

We should be standing on the savannah under the boiling African skies. We should be pacing a slope in the Great Rift Valley, picking out fragments of our ancestors from among the gobbets of lava. We should be in one of palaeoanthropology's frontier outposts, which resound like the names of battles blazoned on regimental colours; Olduvai, Turkana, Krapina, Qafzeh, Zhoukoudian, Jebel Irhoud, Devil's Tower, Teshik Tash. We should be somewhere primal or somewhere heroic, if we want to follow the well-trodden paths of stories about how we got to be who we are.

There are other stories to be told, though, and what counts is not where you start from, but where you go. For this particular excursion, an ideal point of departure is not in the Great Rift Valley, but in the Thames Valley; on the A355 road from Windsor to Beaconsfield, in the shadow of London's south-western quadrant. According to John Wymer, author of a monograph on Lower Palaeolithic archaeology along the course of the Thames, to travel this road is a good way to appreciate the

geological terraces created over the ages by the river's flow. We
don't need to emulate Richard Leakey, bringing his own light
plane in to land on a dirt strip by Lake Turkana. Writing in the
1960s, Wymer recommended taking a number 441 bus. 'The top
deck affords a good view and the continual changing of gears
corresponds with the ascent of the various bluffs between the
terraces.'[1]

These strata are the compressed residues of ancient environments. For archaeologists, their fascination lies in their seasoning
of stone tools, particularly ones from the Lower Palaeolithic, an
era which ended perhaps 200,000 years ago. 'Only France can
qualify as having sites of comparable numbers and richness,'
boasts English Heritage, 'but southern Britain has the additional
merit of having been in closer proximity to the numerous
advances and retreats of the Arctic ice sheets.' The marks of their
passage allow the sites to be placed in a chronological sequence.[2]

The makers of these tools are classified as members of the
genus *Homo*, a branch of the hominid family, whose only extant
member is our own species, *Homo sapiens*. They are sometimes
known as hominines, as distinct from australopithecines, the other
universally accepted hominid genus. Some scholars of human
evolution maintain that the *sapiens* category should embrace the
Lower Palaeolithic hominids; others argue that they should be
classified as different species, and have proposed a variety of candidate names.

The debates between the 'lumpers' and the 'splitters' are not
pedantic wrangles about what to write on specimen labels, but are
fundamental to the study of human origins. They do not, however, need to feature very prominently in this story. What all
scholars agree on is that *Homo* has encompassed great variety
across its span of two million years or more. The makers of the
artefacts that seed the Thames gravel beds were very different
from living humans. They are properly called hominids, but
should they be called humans? Were they people too?

The purpose of the second part of this book is to explore those questions. The moral of the tale is that they were human, but not as we know it. They had minds which were fundamentally human, but profoundly different from ours.

Though it is true to say that these minds lacked some of the capacities of ours, this misses the more interesting point. We cannot imagine what it was like to be an ancient human by trying to think in slow motion. The ancient mind had a different architecture. There is an expression which hovers uneasily between affirmation and euphemism in its contemporary context, but sits a lot more comfortably in the Lower Palaeolithic. Embracing both of the tropes in the phrase, we can usefully describe the ancient hominids as 'differently abled'.

There is no clear window into that difference. The best we have are the scattered panes of ancient artefacts; and flint is opaque. At first glance, Lower Palaeolithic tools may seem too simple – too primitive – to pose any great problems of understanding. But they are infinitely harder to comprehend than the most complex modern machines. A small proportion of people can understand satellites or particle accelerators. Nobody can really understand Palaeolithic handaxes.

Some of these objects, such as the trove found in a celebrated site at Furze Platt in Berkshire, lay where their makers dropped them. In southern England and continental Europe, these are the exceptions. John Wymer explains that the vast majority of 'palaeoliths', artefacts of the Stone Age, were transported by river waters to the sites where they lodged. The rivers ground soft rock to silt and swept it out to sea; the hard rocks were reduced to gravel and deposited in beds along the rivers' courses. In southern England, the only hard rock to speak of is flint, cached in the gravel banks or simply left on the surface after its chalk bedding has weathered away. Wymer calls it 'an almost perfect medium for knapping', the craft of working stone with stone. It is hard and sharp; and yet it co-operates.

Different parts of southern England have different varieties of flint, as English counties have their various cheeses. There is chocolate-brown flint in Surrey, grey in Lincolnshire; the gravels dug at the Kent village of Swanscombe, where pieces of an ancient hominid skull were found, yield a brown and yellow banded strain. It resembles fossilized beeswax, and its characteristic fracture pattern is known locally as the 'Swanscombe eye'. Archaeologists and knappers are connoisseurs of the various strains. 'The finest quality flint for knapping is found in East Anglia; it is black, relatively free of inclusions, wonderfully even in texture and has a slight lustre when freshly broken,' writes Wymer. The worst is the flint of the Chilterns and Berkshire Downs, but good flint can be found in the North Downs, South Downs, Wiltshire Downs and in Dorset.

The quality of a piece of flint depends in part on how long it has been removed from its bedrock. Flint is very susceptible to weathering, which erodes its surface to a tired patina that is often stained by surrounding minerals. The more it has been exposed to the elements, the more it is likely to contain fractures. Modern knappers refer to flint fresh from its chalk matrix as 'green flint', but most of their ancient counterparts would have had to rely upon material exposed to frost or battered by river currents.

Flint's ultimate origins are animal. As the myriad organisms of ancient seas settled into sediment, the calcium carbonate of crustacean shells became chalk, and the minute glassy needles that formed the skeletons of sponges became the nuggets of flint inside the chalk seams. These skeletons were made of silica, which was dissolved and then precipitated, by changes in pressure, into what the British Museum's guide *Flint Implements* calls 'nodules of irregular and often fantastic shape'. They are sometimes arranged in rings, sometimes form flat seams, and sometimes appear as upright hollow barrels, known as paramoudras.[3] Where ancient knappers used other types of rock, these were usually siliceous too: chert, quartzite, the biblical-sounding jasper and chalcedony; even obsidian, a natural glass thrown up by volcanoes.

Modern builders and civil engineers rely just as heavily on silica, in the forms of sand and gravel. To sustain London's great Victorian offensive into the surrounding countryside, quarriers took scores of bites out of the Thames Valley gravel beds. These enterprises continued through the early decades of the twentieth century. After the Second World War, quarrying was mechanized, but before it the quarrymen worked by hand. At only arm's length from their material, they were able to spot the occasional stone curios which they could sell to collectors. One of the most adept of these proxy archaeologists was George Carter, who worked at Cannoncourt Farm Pit, where the 'Furze Platt industry' was unearthed. Wymer relates that Carter was 'passionately fond of gramophone records', and spent most of the money he earned from his finds upon them.

George Carter made his most remarkable discovery in March 1919. It is a slab of flint shaped like an almond and thinned to a narrow edge all around, resembling in colour a bar-room ceiling stained by tobacco. What makes it so remarkable is its size. At 12⅝ inches (321mm) long, it is the largest artefact of its type to have been found in Europe. Even so, it comes with a tale of the ones that got away. According to the wife of the collector who bought it, half a dozen other pieces of similar proportions were discovered close by.

The Furze Platt Giant, now displayed at the Natural History Museum in an exhibition called 'Earth's Treasury', is an example of the type of artefact known as a handaxe. At first glance it's a plain, down-to-earth name for a simple object; but it is just the start of the confusion that surrounds attempts to make sense of such artefacts. It fits the Furze Platt specimen especially poorly. For a start, the Giant does not invite the hand. Like many others of its kind, knapping has left it with scalloped indentations that one can persuade oneself are holds for fingers or thumb. Taking such a cue, one's thumb fits into a scoop near the broader end of one face; this guides the fingers into a supporting position underneath. It's how you might hold a book while walking in the garden, not how you would poise a tool to do some useful work.

You might rightly object that a mind approaching the Giant as a mystery would be likely to find an attitude suitable for contemplating the object's elegant form and bold dimensions. More practical minds might be more adept in finding a practical way to hold it. The Giant's maker would have had a lifetime's experience of handling stone artefacts, and is quite likely to have been robust in build. Nevertheless, the Giant is too large to be a serviceable tool for a hominid much less than twelve feet tall. And however horny the hand that grasped it, a sharp edge all the way round is not the most ergonomic design.

There is nothing exceptional about the Furze Platt Giant's shape – but that is exactly what makes handaxes so extraordinary. They have been found near the southern Cape of Africa, in Palestine, in India, and across much of Europe. The oldest ones known are around a million and a half years old. Specimens from Konso-Gardula in Ethiopia have been placed at 1.4 million years; while recent revisions to the dating of the beds at Olduvai Gorge, in Tanzania, suggest the earliest layers containing handaxes were laid down 1.5 or 1.6 million years ago.[4] The most recent are younger than 75,000 years.

This gives them a unique place in the history of hominid artefacts. As the distinguishing mark of the archaeological record for over a million years, they are central to an understanding of how hominines lived and thought for nearly half their span of existence. If we want to make sense of how we came to have the minds that we do, we can't ignore handaxes.

They are by no means all alike, but the overwhelming thing about them is how little they change across three continents and well over a million years. The basic form is constant: two faces, with a common edge for at least most of their circumference, and a roughly triangular or pear-shaped outline. Almonds, pizza slices, leaves, plectrums, teardrops.

2

Archaeologists assign stone artefacts to 'industrial traditions', named after 'type-sites' where objects deemed typical of the traditions were first investigated. The one to which handaxes belong is called the Acheulean, after Saint-Acheul, a suburb of the northern French town of Amiens. What to call the objects themselves is more problematic. The French archaeologist Gabriel de Mortillet coined the term *coup de poing*, or punch, which is dramatic but ambiguous. 'Handaxes' implies striking blows with an edge, but the force might have been directed through the point, or the edge used to saw and slice. To confuse matters further, in some examples the two ends look as though they have been designed for different purposes. 'Often even the butt of a pointed hand-axe is so much sharpened by trimming that it is difficult to believe it was held in the hand,' remarks the British Museum guide, 'while the point is often so delicate as to seem too fragile for any use which the comparatively massive butt would suit.'

Among the more adventurous proposals are the suggestion that they were thrown at targets, and the claim that they are by-products of the manufacture of other objects altogether. Scientifically speaking, the alternative term 'biface' is preferable to 'handaxe', being more descriptive and making no assumptions. It seems overly reticent, though. The objects are cryptic enough already. Here they will continue to be known as handaxes.

The different varieties of handaxe have been given more fetching labels. Long narrow ones, like spearheads, are known as lanceolates; oval ones are ovates and heart-shaped ones are cordates, tongue-shaped ones are linguates and those that look like almonds are amygdaloid; the name *ficron*, from the French term for a punt pole, is given to ones which have round butts that fit into the hand at one end, and sharply tapering points at the other. Another French coining is *limande*, after the flatfish known in Britain as dab. It's just not true that once you've seen one handaxe, you've seen them all.

Besides shape, materials provide another source of variety. Quartzite axes feel very different from flint ones, denser and coarser to touch. With their stippled surfaces, they look less artificial than worked flint, though imposing a form on them must have taken just as much effort, if not more. Obsidian, with its sheer fracture planes of black glass, seems almost to have a better idea of form than those who worked it.

Archaeologists differ over whether the knappers' designs themselves developed. Glynn Isaac saw the thousand millennia of the Middle Pleistocene as an era of 'restless, minor, directionless changes'.[5] John Gowlett, by contrast, is confident that there was progress in the Acheulean. There is 'no doubt' that awareness of form, variety of tool types, and complexity of manufacturing techniques all 'developed greatly', he has declared.[6] Even so, this is a story of technological change at a geological pace, that took about half *Homo*'s time on Earth.

The handaxe era followed the first million years of hominid toolmaking, which began with a tradition of simple flaking known as the Oldowan (named after Oldoway, as Olduvai was known when the Leakeys first set up camp there). Oldowan tools are irregular in shape, with no real common pattern. But according to the researchers who discovered the oldest known stone artefacts, created in Ethiopia two and a half million years ago, their makers knew the basic principles of knapping. The work of these extremely ancient hominids indicates 'a clear understanding of conchoidal fracture mechanics'. Another collection, assigned the slightly more recent date of two and a third million years, was described by its finders as 'typical Oldowan'.[7]

The form of the handaxe appears quite suddenly, but the earliest specimens seem to retain the Oldowan spirit. Many are massive things, shaped like dogs' heads and similar in size. Others have been knapped just enough to produce a roughly triangular outline and no more. Sketchy as it is, though, the form is there. Unlike the hominids who chipped out the Oldowan tools, the makers of these artefacts appear to have had an idea in their minds of the shape they wanted to produce. They seem to have envisaged a form and imposed it on their material.

It is tempting to see the evolution of the handaxe as a gradual revelation of that form, as knappers become more skilled in realizing it. But the vision shines out even in very old handaxes, such as those found at Olduvai. An edge is carefully worked into a straight edge; a face is shaped into a classic Acheulean teardrop, although the reverse is left coarsely finished. It is difficult to hold such an object, turn it in one's hand, imagine the intentions behind the hammer-strikes, without thinking of the individual who made it as a person.

There are Olduvai handaxes, over a million years old, that look like what the hominids at the Boxgrove site in West Sussex, half a million years ago, would have made if they had been given Olduvai rock. Others look very different, both in shape and degree of refinement. At Olduvai, the Acheulean seems to be exploring its possibilities. At Boxgrove, it has consolidated into a highly standardized form. In the process, it seems to have developed an emphasis on shape and symmetry as well as utility. At the other end of the size range from the Furze Platt Giant are tiny leaves of flint as little as an inch and a half in length. The makers of objects like these seem to have put enormous effort into producing the biggest or the smallest artefacts they could make, rather than the most useful.

Between these extremes, many handaxes are clearly effective tools. Modern knappers have found that they are good for butchering animals. But so are the unmodified flakes knocked off in their preparation, and the irregular edged tools which continued to be made during the handaxe era. To their surprise, the energetic experimental knappers Kathy Schick and Ray Dezzani had no difficulty slicing through inch-thick elephant hide with flakes they struck while making simple Oldowan-style artefacts. As Schick and her colleague Nicholas Toth point out, they were able to butcher 'the world's largest terrestrial mammal with the world's simplest flaked stone technology'.[8] Glynn Isaac listed possible uses of early stone tools: butchery, hacking sticks or clubs from branches, sharpening sticks, opening up beehives, digging into logs to get at larvae, peeling off bark and shredding pith.[9] He concluded that the simplest assemblages of flakes and lumps with jagged edges would have been adequate for all of these tasks. It is questionable whether the improvements in efficiency offered by handaxes were great enough to repay the extra effort required to make them.

A number of thinkers, starting with H. G. Wells in 1899 and including Louis Leakey, have suggested that handaxes were thrown at living targets. It would certainly be remarkable if no hominid had ever flung a handaxe at a prey animal during the million years of the Acheulean, but whether bifaces made good missiles is another matter. The 'projectile capabilities' of an ovate handaxe were investigated by an undergraduate named Eileen O'Brien, possibly with tongue in cheek. She had a replica cast in fibreglass, with weight added to approximate that of the original, just under 2kg. The specimen came from the Kenyan site of

Olorgesailie, which boasts an abundance of remarkably large bifaces.[10] Thrown discus-style by two trained athletes, one expert with the discus and the other with the javelin, it ascended spinning horizontally, then turned through ninety degrees as it came down. More than nine times out of ten it landed on edge, and in two-thirds of the throws it impacted point first.

A similar exercise was conducted by William H. Calvin, a neuroscientist, who thought that the flying biface might fit into his unfolding scenario of early hominid predation. He began with the suggestion that hominids might have discovered the effectiveness of throwing branches into herds of large animals gathered at waterholes. Only one individual might be directly injured, but the herd would stampede, thereby trampling it or leaving it exposed. A stone might have a similar effect, Calvin reasoned, but a handaxe would be better. Instead of bouncing off the animal's back, taking a large proportion of its energy with it, it would tend to strike with its edge, roll forward and snag the animal's skin with its point, imparting most of its force before dropping to the ground. Animals struck on the back tend to bend their rear legs reflexively: Calvin suggested that the particularly painful effect of the snagging point would increase the likelihood that this reflex would be triggered. He called this 'neurological knockdown', while the general idea that handaxes were spinning projectiles has come to be known as the 'killer Frisbee hypothesis'.[11]

Although Calvin asserts that a handaxe's shape would allow it to punch well above its weight, the smaller bifaces do not fit credibly into his scenario. He is also obliged to discount axes without edges all the way round as ones which failed to make the grade as projectiles, and so were used in other ways. Yet these are the sort which lend themselves quite readily to the most straightforward explanation, that they were made to be general-purpose hand tools. The one thing that the killer Frisbee hypothesis does very satisfactorily is illustrate how far scientists will go in search of an explanation for the handaxe.

Iain Davidson and William Noble, an archaeologist and a psychologist, have responded to the Acheulean conundrum by denying that handaxes existed. They argue that archaeologists have misled themselves by seeing bifaces through modern eyes. Although a biface looks like the goal of a manufacturing procedure to us, Davidson and Noble argue, it was merely a source of raw material to an Acheulean. The purpose of knapping was to obtain sharp-edged flakes, and the bifaces were just the residual pieces of rock. They may have been carried from place to place, a portable source of new flakes, and they may themselves have been used as tools, but the flakes were to the fore, and the bifaces were background objects.

The bifacial shape, Davidson and Noble suggest, is no more than the result of the application of a limited repertoire of knapping techniques to material which imposed its own constraints on form.[12] They cite Bruce Bradley and C. Garth Sampson, who discovered, when trying to replicate bifaces found at the British site of Caddington, that a typical Caddington handaxe form invariably appeared at some stage in the process.[13] But the local style was fairly rough and ready. Davidson and Noble's argument loses persuasiveness when extended across the range of handaxes. It takes a modern knapper a great deal of effort to replicate the more highly worked types. There are only so many ways to knap stone: if Acheulean forms were the outcome of constraints, it seems likely that knappers would have discovered techniques which simulated these constraints, and thus reduced the amount of conscious effort required.

Among modern humans, designs without obvious function are taken to have symbolic significance. But there is no evidence that hominids who made handaxes had any symbolic capacity. They have left behind no paintings, carvings, personal adornments or graves. They just show a predilection for the almond-shaped form, and hints of an appreciation of its symmetry.

3

'They're biofacts,' said the Boxgrove archaeologist, holding a biface in one hand and miming the action of striking it with a knapping hammer. The term 'geofact' is used to denote a stone which looks like an artefact, but owes its appearance to geological action. He was implying that handaxes were made by instinct rather than art; and compared them to the elaborate structures constructed by bowerbirds, which also appear to show an appreciation of form and symmetry. These are textbook examples of sexual selection: their extravagance and intricacy signal that their makers are fit individuals, able to undertake demanding activities, and are therefore good potential mates. Maybe female hominids only mated with males who demonstrated competence in handaxe-making, he suggested.

He didn't mean it seriously, but his offhand remark immediately suggested to me a new way of looking at handaxes. For the best part of a century after Darwin introduced the concept of sexual selection, it lay neglected. It took the new wave of Darwinian theory, beginning in the 1960s, to make sense of sexual selection. Now it is at the heart of a new understanding of evolution, based on the recognition that males and females have different reproductive interests.

Handaxes need something that elemental to account for them. For the first time in hominid history, an artefact form is standardized. It is reproduced over an immense geographical range, for a period of time unrivalled by any other known artefact. Some great force must have perpetuated the handaxe form and held it constant. Culture as we know it did not exist. What else was there, but sex?

The minds that made Acheulean handaxes were not bowerbird brains, though, needing little more comprehension of their actions than does the loom that weaves the cloth of gold. Some scholars have inclined to the idea of a dedicated program, however, suspecting that the handaxe standard was maintained by hard-wired circuitry. The Boxgrove fieldworker echoes this suspicion in the term 'biofact', and when he goes on to suggest that the handaxe form may have arisen from a particular kind of neural organization in the Acheuleans' brains. Something about their cognitive processes – circuits configured to detect symmetry, perhaps – caused the handaxe shape to 'fall out' of stoneworking activity.

We're standing in what looks like an old livestock shed. (The Boxgrove diggers, mostly students, eat their meals in the one next to it.) Ranks of handaxes are laid out on tables, displaying their essential unity of form. To the modern eye, surveying it with an aesthetic sense, this is a fine collection. The bifaces are mostly ovate, sometimes delicate, and often assiduously worked. They look like the results of a detailed plan.

Davidson and Noble, however, propose just the opposite. Scientists speak of results 'falling out' when they mean that the configuration of a system determines a particular outcome. If Davidson and Noble are right, handaxes fell out of flake production activities. Since the end results their makers sought were the flakes, not the biface remainder, they would have had no need to think beyond the flake they were striking. There was no plan, just a repetitive procedure sustained as long as the material allowed.

To some extent this argument is a reaction against an archaeological tradition which applies contemporary Western aesthetic standards to ancient artefacts. The early collections are biased towards handaxes, because those were what the collectors wanted, and the men at the quarry face were not trained to pick out less obvious artefacts. Davidson and Noble argue that archaeologists tend to think of stone tool history as a narrative of progress, with chopper tools replaced by handaxes, which were in turn replaced by more sophisticated technologies. In fact, the story is more one of accumulation, with new techniques being added to older ones, rather than supplanting them.

The conventional assumption, however, is that handaxes represent the intended results of the knapping operations that produced them. If so, their makers must have had some sort of mental model from which to work. Elements of that model may have fallen out of their cognitive apparatus, but those axes did not just fall out of the raw flint. The Boxgrove handaxes bespeak considerable skill, and dedication far beyond the demands of utility.

We can't take it for granted that we can do anything our ancient forebears could. The Neanderthals, the immediate predecessors of modern humans in Europe, made tools using a method known as the Levallois technique, which involved preparing a stone core and striking flakes off it. In a spirited defence of Neanderthal capabilities, Brian Hayden suggests that fewer than a score of modern knappers have mastered the technique. He observes that it seems far more difficult than core techniques developed much later by modern humans.[14]

Handaxes and other stone artefacts are easier, but demanding none the less. Phil Harding is one of the best flint-knappers in Britain, and he reckons it takes about six months to learn the basics of the craft. He has been practising it for about twenty-five years, inspired by a 'passion for flint'. A native of Wiltshire, he secured his first archaeological job at the Norfolk site of Grimes Graves, where some of the finest flint in the land was excavated by Neolithic miners some 4,000 years ago. Today, he lives in Salisbury and works for Wessex Archaeology.

I visited Phil to watch flint-knapping at first hand, and equally importantly, to hear how a modern mind puts the task into words. Outside his small terraced house stood the pick-up truck that he uses to transport his raw material; outside the railway station, a direction sign for the nearby stone circle of Stonehenge served as a reminder that around these parts, there's quite a precedent for bringing rock from far away.

The first thing Phil did, before stepping out into his back garden, was to pull on a pair of old boots, almost fossilized themselves. These were protection, he explained. Once, a piece of flint had sheared off from the rock he was breaking, cutting deep into his foot. He works sitting upright, wearing plastic goggles, and a piece of leather as an apron. Although the ancient knappers would have lacked artificial protection for feet and eyes, he is amused when he sees pictures of entirely naked hominids breaking flint without anything to cover their laps.

Phil set his chair down next to a heap of flint, mostly chunks of rock the size of loaves or larger, piled up by his garden shed. This was a haul he had gathered at Boxgrove a few weeks before. I was going to witness the making of a Boxgrove handaxe.

First Phil demonstrated the rules of knapping; both of them. There are a few accompanying technical terms, the two basic ones being 'flake', which is the piece struck off from the rock being knapped, and 'core', which is what remains. (The accumulated flakes are called the *débitage*.) The spot where the blow is delivered is known as the 'striking platform'; the energy of the blow travels through the rock to produce a characteristic fracture pattern of concentric circles. This pattern, known as 'conchoidal' or shell-like, is the key to the workability of the stone.

When the shockwave reaches another surface of the nodule, the flake detaches. Imagine a piece of flint the shape of a roasting tin, or an ingot, wider at the top than the bottom. The angle between the top and side is less than ninety degrees. If the top is struck just inside the edge, the shockwave should travel downwards until it meets the side surface, detaching a wedge-shaped flake. The first rule of flint-knapping is that the angle should not exceed ninety degrees, but should approach it, ideally falling between seventy and eighty degrees. And that is the essence of knapping. Like the theory of natural selection, though, it gets a lot more complicated in practice.

The edge of the flake scar makes a new angle with the rest of the core, creating a ridge. Phil struck another blow, and pointed out that the ridge now ran along the back of the resulting flake's freshly exposed surface. That's the second rule of knapping: flakes follow ridges. The tighter the ridge, the more elongated the flake. Add this to the first rule, and a modern human is ready to knap flint.

So Phil sat and knapped, while I stood filming him with a camcorder. At this particular point in history the video camera stands for technological sophistication in a way that other objects in the scene do not; the trains which passed the garden every so often, for example, or the boots on our feet. But if the technological distance from Phil's workboots to my camcorder is like the distance from the Moon to Earth, the Boxgrove hominids' handaxes are as far away as Saturn.

Like the earliest handaxe-makers, a million years before them, the Boxgrove knappers could have reduced their workload by choosing rocks that already bore some resemblance to the final result. Phil picked out a roughly triangular slab and indicated how its shape could be refined into a handaxe shape. Roughing out such a piece may take five minutes or so, by which point the knapper already has a usable, if crude, edged tool. But Phil decided to go for a nodule that presented a challenge. He selected a crusty, bubonic block the size of a car battery and devoid of any trace of regularity.

Turning it in his hands to find a way in, he began to reduce it using a round hammerstone. It takes brawn to send shocks through flint with sufficient force to split great plates of stone away. To create form in the process requires brain and a shrewd eye as well. Once the handaxe shape had begun to emerge, Phil switched to a piece of antler for the lengthy process of crafting the edges. Knappers call such tools 'soft hammers', and use them when delicate flakes have to be struck.

After forty-five minutes' work, some awkward patches when the flint wouldn't break the way it should, a few spots of blood and a scattering of sharp flake debris, Phil had completed a hand-axe. It was a little more than ten centimetres at its longest, after he had knocked off the tip to illustrate what is known as a tranchet edge, and about ten centimetres (four inches or so) at its widest. Ovate in shape, it would have fitted in nicely with the specimens laid out on the table at Boxgrove.

Its shape was not determined by the rock from which it was struck, Phil commented, but the material does tend to influence the form. Making a pointed biface would have been more difficult because of the nature of the Boxgrove flint. Too 'blobby', he said, though he allowed that it was good quality flint, nice and fresh. An academic paper on Boxgrove handaxes notes, however, that the local rock is 'full of chert-like inclusions'; these make the texture of flint less even.[15]

This handaxe may have been pushed in a certain direction by its origins, but it scarcely fell out of the rock. It was produced by sustained effort, careful consideration before each blow was struck, a certain amount of doggedness, and a mental model of the general form it should take when finished. Phil Harding is one of the few really skilful knappers in the country, and he had to work hard – with material chosen for its difficulty, admittedly – to pro-duce a common or garden handaxe.

There's another modern handaxe on display in the British Museum, the work of Mark Newcomer, another highly accomplished knapper. Exquisitely symmetrical from the steady arc of its base to its surehanded point, it is a handaxe fashioned with a jeweller's eye. It is made from silky charcoal-grey flint; the flake *débitage* has been reassembled and the axe mounted at the centre of the reconstituted core, like a satin teardrop in an Easter egg. Without any impression of the flinty realities of the knapping process, it would be easy to imagine that the shape was predestined, to be extracted by some subtle Platonic craft. When I go and look at it now, though, I imagine a great pile of *débitage*, representing earlier attempts that did not reach this summit of excellence.

Having seen how a modern human makes a handaxe, I am all the more sceptical about the proposition that Acheulean bifaces were merely residues of flake production routines. But handaxes were made by Acheulean *Homo*, not modern *Homo sapiens*. To make any sense of the behaviour of these ancient hominids, they must be considered as figures in a landscape.

4

A great chalk cliff is the backdrop, seventy metres high, a short section of palisades twenty miles long in all. Unrolled in front of it is a pebbly beach, protected from the sea by an outstretched arm of land, which also encloses a tidal lagoon. There is a freshwater pool fed by a stream, a stripe of flint at the base of the cliff, a thatch of woodland at its top; a notch, shallow enough to climb, carved into it by a spring. The broader picture is of a mosaic of grassland, beach, fresh and saltwater pools, homogenizing into mudflats as it approaches the sea.[16] Half a million years ago, between two icy phases, the climate is much like today's.

The hominids probably lived in the more hospitable woodland at the top of the cliff, making sorties on to the beach to obtain meat. Many animal bones have been found at the site, but nearly all of those bearing cutmarks come from larger animals, suggesting that the hominids disregarded small herbivores in favour of species such as red deer, giant deer, horse and rhinoceros. The herbivores themselves may have been attracted to the beach because it was a source of salt, or because it served as a corridor bypassing the woodland.

For Clive Gamble, of Southampton University's archaeology department, the Lower Palaeolithic archaeological record testifies to a lack of cohesion among the hominids it represents. There are no hearths, no 'conversation rings' of debris scattered in circles, the 'archaeological signature' of spoken language. Gamble believes that the hominids communicated with each other by gesture, and lived highly individualistic lives. Their projects were solitary ones: they made handaxes, not huts.[17]

In these respects, the Boxgrove site is no different from others of the period. But Mark Roberts, Director of the Boxgrove Project, thinks that social activities took place backstage. He envisages the hominids butchering carcasses on the beach, then carrying the meat back up to camps in the woods. They would have worn very simple clothing, he suggests, noting that one of the handaxes shows a wear pattern characteristic of hide scraping. But even the simplest wraps would have required a subtle form of knowledge, since hide must be cured to prevent it from either going rigid or rotting. Roberts also believes that the acquisition and processing of meat depended on a level of co-operation which would not have been possible without language.[18]

However differently they interpret it, though, archaeologists agree that Boxgrove is an exceptional site. Almost all the other Lower Palaeolithic sites in southern Britain are assemblages of objects carried by water from other locations. Boxgrove is a preserved landscape, the largest of its kind in Europe. It allows us to see something of the relationship between handaxes and other elements in the Acheuleans' world, the terrain in which they were deployed, and the animals upon which they were used. Boxgrove is a picture window into the hominid behavioural ecology of half a million years ago; not a window into the early hominid mind itself, but the best available platform upon which to build a model of that mind.

Today the sea has retreated several miles to the south, the original cliff long since crumbled on to the shore. A cliff has been recreated by human industry, however; first that of the Amey company's quarrying operation, then that of the archaeologists. Their excavations have something of an Ancient Egyptian style: an ascending terraced slope, like the side of a pyramid, overlooking a level grey chalk floor in which rows of rectangular holes are carved, resembling cisterns. These are known as trenches, and the military echoes extend to the living quarters in the old farm buildings, like a commandeered billet to the rear of the line. The archaeologists bear little resemblance to the subaltern class of the Great War, though. They look a lot more like the kind of young person found at the protest camps, also sometimes featuring excavations, that spring up in the paths of roadbuilding projects.

Or rather, they did. The last digging season at Boxgrove was in of 1996. Like road protesters, the dig volunteers and their supervisors were followed by earthmoving machinery. A fifteen-foot layer of sand and soil was spread over parts of the site, to seal it in for posterity.

After ten years, it was time to take stock and publish the results. As well as the 400 handaxes, there was a wealth of *débitage*, and around twenty flakes that have been 'retouched' in ways that make them useful as scraping tools. There were also two soft hammers made from deer antlers. These tools were evidently used and looked after for a long period. They may be the oldest such 'curated' objects in the archaeological record.

While Lower Palaeolithic artefacts are scattered liberally across southern England, human remains from this period have been found at just two sites in the region. One is Swanscombe, where two pieces of the same skull were discovered in 1935 and 1936, followed by a third in 1955. The other is Boxgrove, where a length of shinbone was unearthed in 1993, followed by a pair of lower incisor teeth two years later. The Boxgrove investigators have designated 'Boxgrove Man' as *Homo heidelbergensis*, a species named after a jawbone found at Mauer, near Heidelberg, in 1907. Such hominids were different in various respects from modern humans, but their brains were of modern proportions. Traditionally, they have been considered not to be a separate species from our own, but an archaic form of *Homo sapiens*.

5

Hominid taxonomy is a restless and quarrelsome business. Its instability arises from the combination of meagre samples and a range of variability which is indisputably large, but whose pattern is endlessly debatable. Apes and humans are classed as members of the hominoid superfamily. All the species on the branch that splits from the apes and leads to humans are members of the hominid family, while the apes have traditionally been sorted into families of their own. Newer methods of assessing relationships, however, encourage Jared Diamond's view that chimpanzees and gorillas should be classed as hominids.[19] In other words, African apes belong in the same family, whether they are naked or hairy.

Thankfully that's a question which doesn't bear on this story, which follows the lineages on the human side of the hominoid divide. The last common ancestor of modern apes and humans is believed to have lived between eight and five million years ago. Fossils which may date from the more recent end of that range have been found in Ethiopia, but their status is undecided at

present. The oldest named hominid fossils are seventeen fragments 4.4 million years old, from the Ethiopian site of Afar. They
were first given the designation *Australopithecus ramidus*, though it
has since been argued that they should be put in a genus of their
own, called *Ardipithecus*. Then comes another collection of fragments, dated to about four million years ago, and designated
Australopithecus anamensis in 1995. The best-known taxon is
Australopithecus afarensis, now most fully represented by the partial
skeleton known as Lucy, three million years old.

Australopithecines distinguished themselves unequivocally
from apes by adopting an upright posture. We do not have to rely
solely on inference from the shape of their fossilized bones to
know this, because we also have some of their footprints. A trail
was left in volcanic ash by two or three australopithecines 3.6 million years ago, and found by Mary Leakey's team in 1979, at
Laetoli in Tanzania.[20]

The earliest known australopithecines were small creatures
with brain sizes in the range of 400 to 500cm^3, similar to those of
modern apes, and are often thumbnailed as 'upright chimpanzees'. They radiated into a variety of forms. One species,
Australopithecus africanus, appears to have rediscovered the advantages of life in trees. Whereas the skulls and teeth of early
hominids are rather traditional in the support they give to the
conceit that evolution is progressive, displaying an orderly trend
towards modern forms, the rest of the *africanus* skeleton gives a
puckish twist to a story that was too neat to be true. During the
apartheid era, fossils gathered by the South African palaeoanthropologist Phillip Tobias remained little known outside the country.
Studies in the late 1990s on some of these specimens, from a
limestone quarry at Sterkfontein, have shown that while *africanus*
continued hominid progress towards modernity above the neck,
the rest of it took a step back towards the ancestral condition.
With its long arms and short legs, it resembled apes in its proportions, and seems to have spent more of its time in trees than did

earlier australopithecine species. In this case, evolution went upward rather than onward.[21]

Others stayed firmly on the ground and grew highly robust as well; the last of these, also known as paranthropines, died out about a million years ago. Until 1995, the australopithecines were thought to have lived along the diagonal slash of the Great Rift Valley, adapting to the savannah on its eastern side, while apes continued to live in the woodlands on its west. Then, in one of those frequent puffs of conjuror's smoke that throw the cards of the human origins pack up in the air again, a piece of australopithecine jaw turned up in the Central African country of Chad. Its finders have assigned it to a new taxon, *Australopithecus bahrelghazalia*. Others consider it should be classed as *afarensis*, and one can see their point of view.

Some of the earliest specimens assigned to the genus *Homo* have been estimated to be 2.4 million years old, though scientists generally make their calculations from a cluster of fossils 1.9 million years old.[22] Early *Homo* is a motley group. It includes fossils with cranial capacities of only half a litre, smaller than the largest australopithecine cranial volumes, while others are past the three-quarters mark. A number of specimens claimed for the genus have since been reclassified as australopithecines. Of the remaining ones, some have bodies with apish undertones and heads that are hominine in character, while others have hominine bodies combined with teeth and faces that resemble those of robust australopithecines. They used to be lumped together as *Homo habilis*, though palaeoanthropologists have never found this classification entirely satisfactory.[23]

Between 1.8 and about 1.5 million years ago, another group of hominids displayed forms which scientists readily compare to our own. They have a characteristically human body type, fully adapted to upright motion; they are larger than australopithecines, but have smaller teeth. Their brains are both greater in actual volume, and greater in size relative to their bodies.

The best preserved is the nearly complete skeleton known as the Nariokotome or Turkana Boy, which is 1.6 million years old, and was discovered in 1984 by Richard Leakey's team in the Lake Turkana region of Kenya. Leakey's readiness to identify with it is also strikingly complete. 'The Turkana Boy had been part of a major shift in human evolution,' he has written, 'one in which the seeds of the humanness we feel within ourselves today were firmly planted . . . It was, I believe, at the real beginning of the burgeoning of compassion, morality, and conscious awareness that today we cherish as marks of humanity.'[24]

Leakey would prefer that the Turkana boy should be included in one all-embracing taxon of *Homo sapiens*, wise man, though hominids of this class have been more widely known as *Homo erectus*, upright man.[25] Lately, scientists have started to use the term *Homo ergaster*, working man, for African members of the group. Some consider that the designation of *Homo erectus* should be confined to the Asian population. Others accept *ergaster* as a designation for early specimens of the African group traditionally classed as *erectus*, but speculate that younger fossils represent a migration of *erectus* from Asia back to the ancestral territory of Africa.

By the Boy's time, it now appears, hominids had diffused out of Africa as far as China. Until recently, the first settlement of Asia was believed to have taken place considerably later, but a tooth from the Chinese province of Sichuan has been redated to 1.9 million years ago, the same vintage as some stone objects, which may be crude tools, found in northern Pakistan. A range of dates from 1.6 to 1.8 million years ago has also been obtained for skull specimens from Java.[26] At the other end of the timeline, revised calculations have given ages of between 53,000 and 27,000 years for the youngest fossils of the same type. The revisions suggest that hominids arrived in Eastern Asia very early, and then spent most of their history largely isolated from other populations. That could explain why the Acheulean did not extend east of India.

There is little evidence of hominids at the other end of the Eurasian land mass until the past million years. Older dates have been claimed from a couple of sites on the periphery of Europe. Splendidly isolated, geographically and chronologically, is a controversial date of 1.8 million years obtained for a jawbone found at Dmanisi, in Georgia. Equally remarkable in its way is the date of 1.4 million years given to handaxes from the Ethiopian site of Ùbeidiya, in Israel. Since this is of the same vintage as the handaxes from Konso-Gardula, the oldest Acheulean assemblage known, it implies that the Acheulean industry materialized like a crop of mushrooms from Ethiopia to the Levant.

Claims of dates around or beyond the million-year mark have been made for several sites nearer the heart of Europe, in France, Italy and Spain, but none of these has yet been firmly established. At the southern Spanish site of Orce, for example, a piece of skull dated at 1.6 million years was claimed by supporters to be hominid, but sceptics considered it to be horse. Tests on fossil traces of albumin protein supported the case for the hominid.[27] The oldest European hominid remains to have been described in detail are a set from Gran Dolina, one of two sites at Atapuerca, in northern Spain. These are said to be 780,000 years old, and their investigators have coined the name *Homo antecessor* for them, to assert that they represent the ancestors of the Heidelberg, Neanderthal and modern human types.[28]

The Boxgrove and Mauer remains – a chewed shinbone, two teeth and a jawbone – are the only known European fossils that fall near the half-million year mark on the hominid timeline. Other European remains assigned to the archaic *Homo sapiens* or *Homo heidelbergensis* categories are somewhat younger; 200,000 or 300,000 years old. Some scholars consider them to have cousins in Africa, but that is a moot point, like most others in this field.

6

Enough name-dropping. It's time to make a story out of the trail of Latin neologisms, decimal points and millennia. That involves offering some suggestions about why hominids took the forms they did, as well as about when they did so, and what they should be called. For the sake of simplicity, the hominid varieties can be divided into three groups. Into the first go the australopithecines and the earliest kinds of *Homo*, all with brains well under a litre in volume, and bodies still somewhat equivocal about their hominid characteristics. Into the second go *Homo ergaster* and all the types that emerge subsequently, up to the forms known as archaic *Homo sapiens*. The middle of their brain-size range is around the one-litre mark, and their bodies have acquired the distinctive form of the *Homo* genus. The final group comprises the Neanderthals and modern *Homo sapiens*; the former is built much more heavily than the latter, but both have a typical brain volume of around 1.3 litres. With a reasonable degree of correspondence to their anatomies and apparent mental capacities, we can refer to the three groups as the lowbrow, the middlebrow and the highbrow.

This casual scheme demonstrates why it is so important to spend time on handaxes. The middlebrow period is harder to visualize and easier to ignore than the others. Lowbrow hominids are relatively easy to imagine, because chimpanzees can be used as stand-ins. Highbrow ones get most of the attention because they include us and they do the most interesting things, apparently enjoying a monopoly on symbolic culture. But how did hominids go from a para-chimpanzee condition to here? The answer lies in the adaptations they made during the middlebrow phase, of which the handaxe is the signature.

We'll need a closer focus at times, however. For this I'll favour some taxa over others, but without prejudice, as the lawyers say. If I seem to be leaning towards the splitters, separating hominids into species rather than subspecies, it's largely because that seems to offer a better purchase on hominid variation for those of us who have not devoted an academic lifetime to its study. From those who have, I beg indulgence for using the lowest-level taxonomic name *heidelbergensis* without a preface, so as to fudge the question of whether they are a species, such as *Homo heidelbergensis*, or a subspecies, *Homo sapiens heidelbergensis*. Give or take a taxon or two, my hominid *dramatis speciae* runs like this:

- Australopithecines, lumping all the gracile varieties together

- Paranthropines, the various robust australopithecine types

- Early *Homo*, covering specimens designated as *habilis*, *rudolfensis*, or just '*Homo* sp.', meaning 'an indeterminate species of *Homo*'

- *Homo ergaster*, the African species identified from early examples of the type traditionally known as *Homo erectus*

- *Homo erectus*, *ergaster*'s Asian opposite number

- *heidelbergensis*, a type identified to denote some early European specimens, including those of Boxgrove; part of a wide and poorly defined class often known as archaic *Homo sapiens*

- modern *Homo sapiens*, or people like us.

Despite granting myself a writer's licence, the choice hasn't been easy. I have based my cast list on a scheme, drawn up by the palaeoanthropologist Bernard Wood, which commands scholarly respect. It also reflects recent discoveries, which is important because the threat of the future is almost as acute in human evolutionary studies as it is in the computer market. The most important recent initiative is the splitting of the type known as *Homo erectus* into two, with all or part of the African stock renamed *Homo ergaster*. While it will be a long time, if ever, before the revision is accepted throughout the university departments, encyclopedias and textbooks, I have acknowledged it because it is significant to the plot of this story.

About five million years ago, the world's climate descended fitfully into a colder and drier phase. The carpet of forest that covered East Africa began to fray, and holes appeared in the fabric. Filled with grasses, which are less thirsty than trees, these were the spaces in which the hominids arose. Other primates that moved into open country, such as baboons, did so on all fours. This seems like a more straightforward approach, and scientists have long puzzled over why hominids chose bipedalism. One idea, once popular, was that it allowed them to see over the top of tall savannah grasses. The notion is derided these days. If a primate wanted to see over tall grass, anthropologists ask, why would it need to develop a whole new means of travel? Why couldn't it just stand up now and again?[29] That works for meerkats, after all. A more credible argument, developed by Peter Wheeler of Liverpool University, is that standing up straight reduced exposure to the savannah sun.

While this may have turned out to be an advantage of bipedality, it was probably not the reason why hominids stood up on their

hind limbs. Robert Foley, of Cambridge University, sees bipedalism not as a choice, but as a consequence of the way the
proto-hominids were. Baboons and other monkeys that occupy
open terrain are descended from species which ran along branches
on all fours. Hominids were descended from apes, which climbed
or swung through trees, and whose anatomy was more oriented
towards the vertical. An animal can travel across the ground efficiently on either four legs or two. Hominids used two limbs
because changing to four would have entailed a greater reorganization of their anatomy.[30]

They did not complete the reorganization for a long time,
either. Australopithecines retained a body form suitable for animals that spent part of their time in the trees and part on the
ground. From the ground, they could stretch up to take fruit from
the branches of small trees; they could also pick up the food that
lay at their feet.[31] They emerged as borderline creatures, making
the margins between woodland and savannah their own.

Small bipeds glimpsed on the edge of woods are commonplace
in folklore, and australopithecines resembled elves in two respects
besides habitat. One was stature. At a metre in height, some of the
australopithecines fell into the range traditionally reported for
elves, between three and four feet. Others were half as tall again.
One of the many heated australopithecine debates concerns
whether these represent more than one species. The alternative
explanation is that the big ones were males and the small ones
were females. Similar disparities are seen in living species, and are
considered to arise from the way in which these species play the
mating game. Male gorillas weigh twice as much as females, for
example, and gorillas typically form groups in which a single dominant male exercises a reproductive monopoly over what
primatologists call a 'harem'.

Monogamy is another option for primates, but australopithecines do not seem to have taken it. At the Ethiopian site of
Hadar, thirteen australopithecines appear to have died together,

implying that they had lived in a sizeable group. This would represent another significant trait that australopithecines share with elves, which belong to the class of folkloric creatures known as the 'trooping fairies'. By contrast goblins are solitary, as are orangutans, and monogamous primates live in nuclear families.[32]

Other primate species, including chimpanzees, have more dynamic arrangements in which a number of males live in a group and compete for females. The fossil record does not indicate whether australopithecines lived in single-male or multi-male groups, since the anatomical reflection of the difference lies in the size of the testicles. To achieve control over a harem, a male gorilla has to establish dominance over rival males. This puts a premium on size and physical strength, but since reproductive competition ends with these contests, male gorillas do not need to invest heavily in their gonads. Small testicles will suffice, so small testicles are what gorillas have. Female chimpanzees, on the other hand, may mate with more than one male. This places a premium on sperm, and thus male chimpanzees have large testicles to outdo each other in 'sperm competition'.

Scientists do not imagine that species adopt particular strategies through random quirks of temperament. Reproductive strategies are seen as responses to environmental circumstances. Robert Foley and Phyllis Lee explore these responses in the light of the axiom that males distribute themselves in relation to females, while females distribute themselves in relation to resources.[33]

This principle is an expression of the most powerful idea in modern evolutionary theory, that males and females have different reproductive interests. Among mammals, the number of descendants a male leaves is limited by the number of females with whom he can mate. The number of descendants a female can leave is limited by the number of times she can become pregnant. Human males could, in theory, sire thousands of children. A contender for the world record is a Moroccan potentate known as

Moulay Ismail the Bloodthirsty, whose claimed tally of 888 children is quoted with relish by sociobiologists. The women's record is sixty-nine, and that seems the more remarkable score. It is held by a Mrs Vasilyev, a Stakhanovite of reproduction *avant la lettre*, who achieved the feat in four batches of four, seven of three, and sixteen of two. Even more astonishingly, especially in eighteenth-century Russian peasant society, all but two of the children survived infancy.

That record is out of sight of the competition. In tribal societies, men may father dozens of offspring, but women do not raise more than ten children or so. To maximize reproductive success, a male should seek to mate with as many females as possible. A female's reproductive interests, by contrast, are best furthered by securing the food necessary to nurture offspring from conception to maturity.

Sources of food may be spread evenly across the landscape, or they may be patchy. If they are evenly spread, and high in quality, a number of females can live off a single small patch. A single male may be able to secure a group of this kind as a harem. This is what has happened among gorillas, living in forests. But since the australopithecines' environment was becoming drier, Foley and Lee argue, its resources would have been getting patchier, and the clumps themselves would have offered slim pickings. This would have discouraged the formation of female groups, because small patches could not sustain them. Female australopithecines would have been thin on the ground. They would have had to range more widely to get the food they needed, and so males would have had to go farther to find them.

One of the pillars of modern Darwinism is the insight that individuals share reproductive interests to the extent that they share genes. With females dispersed over a wide territory, it would be in the reproductive interests of related males to form coalitions to keep other males out of the territory. This is, in fact, what chimpanzees do. Foley and Lee's case is that the ecological conditions of australopithecine evolution were simply wrong for the development of harem systems. Australopithecine social structure, say Foley and Lee, resembled that of chimpanzees.

If australopithecines were upright chimps in the social sense as well as the anatomical one, this would have implications for the rest of our lineage, ourselves included. Social space, in Phyllis Lee's phrasing, is finite. There are only so many social systems a species can choose. Individuals can associate with related members of their own sex, or with unrelated ones, or avoid contact with their own sex altogether. Contacts between the sexes may be brief encounters, or long associations. These give a total of eighteen basic permutations, though Foley and Lee allow for other variables, and arrive at a maximum of thirty-two possible social systems.

Although primates have widely differing ways of life, from the communally erotic bonobos to the solitary orang-utans, not all of the possible systems are represented, suggesting that some of them are not stable in practice. Some systems are markedly commoner than others, too. Bonds between female kin are the most common basis of social structure, followed by monogamy. The rarest is male kin-bonding. This, however, is characteristic of our nearest relatives, the chimpanzees. It seems that somewhere back down our lineage, our common ancestors replaced the female bonds of their society with male ones.

Once such a shift has taken place, it may be hard to reverse. One adaptation constrains the next, since evolution proceeds by tiny changes to existing structures. Each of these changes must be either neutral or immediately beneficial. The evolutionary biologist Sewall Wright conceived of fitness as an undulating surface formed from points of different heights, representing gene frequencies with different levels of fitness. Natural selection will push a species up the nearest peak in this landscape. It has no vision; it cannot see across the valleys to a higher peak in the distance, and so will not lead the species into a valley of lower fitness on the way to sunny uplands.

Unable to travel through the valleys, primates have a limited number of paths they can take from one social system to another. It is difficult, for example, to move from the kind of system seen among chimpanzees to the kind seen in gorillas. In the denser central forest, not critically damaged by the change of climate, gorilla females could continue to congregate. On the margins, female bonding was not viable, so the option of change in this direction was closed. However, the case of the bonobo shows that the quality of life within the chimpanzee social system can be impressively modified. *Pan paniscus*, the bonobo or pygmy chimpanzee, has found a way of reducing the level of conflict generated by *Pan troglodytes*, the species familiarly but now inaccurately known as the common chimpanzee. Bonobos famously make love not war. Probably the most significant of their prolific sexual engagements, socially speaking, are those between females, which appear to help female *paniscus* develop much closer bonds with each other than female *troglodytes* can manage. This makes female influence much stronger in bonobo society, which is much calmer because of it. Male bonobos remain male chimps, but they tend to bicker and scrap rather than terrorize and kill.

Not all authorities have reached the same conclusions about the evolution of hominid mating systems. Birgitta Sillén-Tullberg and Anders Møller suggest that hominids went through a phase of polygyny, in which females mated with a single male, as among gorillas, rather than with many, as chimpanzees do.[34] Foley and Lee's model receives support, however, from Henry M. McHenry's analysis of sexual differences among hominids. The gross indicator of difference is weight, estimated in fossils from the size of leg joints. In living primates, the males and females of monogamous species are typically the same size. Males of harem species, such as gorillas, are twice the size of females. Male chimpanzees are about 50 per cent heavier than females, and McHenry finds similar differences among australopithecines.[35]

Another mark of male competition is canine tooth size. In australopithecines, the range of canine size is smaller than in common chimpanzees, and is more similar to those seen in bonobos. Then again, australopithecine arms differ sharply in length. As McHenry points out, Darwin had envisaged a shift in hominid combat from the use of teeth to the use of arms, wielding weapons. This would, Darwin observed, have led to a reduction in canine size. But even with modern technology, teeth still count in contests between males, as the staff on the Saturday night shift in a hospital emergency department will confirm.

The balance of sex differences seen in australopithecines, combining large differences in body size with small differences in canine size, is an unusual one. J. Michael Plavcan and Carel P. van Schaik suggest that it is partly to blame for the range of mating systems proposed for australopithecines, which encompasses almost all the options employed by living primates. Reviewing all the possible explanations, they conclude that none is entirely satisfactory.[36]

Our best sketch, then, depicts australopithecines as animals that resembled chimps in important ways, but were distinguished by the adaptations they made to a mosaic habitat of woods and grasslands. Later, as their lands became more arid, their emphasis shifted further towards the open country. Their males were probably competitive amongst themselves, but formed coalitions based on kinship. Their females would have tended to keep themselves to themselves. They combined a substantial degree of intelligence, comparable to that of living great apes, with a new and decisive shift towards bipedality.

Raymond Dart, the scientist who identified the genus, became convinced that they were carnivores who lived by the use of weapons. His ideas were popularized by Robert Ardrey, author of the immensely successful *African Genesis*. In this reading, a myth of original sin for the aftermath of world war, australopithecines were the beast that is now within us. They were, and humans in essence remained, 'killer apes'.

They used to be called 'ape-men', too, implying that they were unsatisfactory versions of both. 'Upright chimps' says a great deal in two words about their intelligence and social structure, but less about their behavioural ecology. If Dart had been a different sort of romantic, or the current academic style of whimsy was slightly differently tuned, the hominids that appeared on the woodland margins might instructively be known as ape-elves.

7

Evolutionists generally consider it the height of gaucheness to talk about progress. They acknowledge that evolution is not a project whose goal is to produce humans, but an accumulation of blind changes, shaped by the pressures of natural selection. Even within the hominid lineage, it is no longer fashionable to talk of the ascent of Man, even though the narrative of brain expansion and the emergence of culture insistently imply it. Taxonomical 'splitters' emphasize the point by imagining hominids not on the rungs of a ladder, with *sapiens* at the top, but as fruits on a hominid bush. During most of hominid history, according to this view, there were several fruits on the bush at any one time. The present state of affairs, in which there is only one hominid species alive, is the exception rather than the norm.

The theoretical equivalent of the bush is the idea of adaptive radiation. A new type of organism appears; it diversifies into subtypes which specialize themselves into different ecological niches. This happened with the australopithecines. Some of them developed large back teeth and immensely powerful jaws, while others

retained their light, or gracile, build. One of the robust australopithecine species had a prominent ridge on the top of its skull,
like a conquistador's helmet, upon which to anchor its massive jaw
muscles. These striking features probably enabled them to eat
foods that came in hard cases, like nuts or tough-skinned fruits,
which ground their teeth flat. A famous specimen discovered by
Mary Leakey was nicknamed 'Nutcracker Man', which may not
have been far from the truth.

The robust australopithecines, or paranthropines, have no
descendants. Looking at their tugboat jaws and millstone teeth,
not to mention those crowning girders, it is easy to imagine that all
this heavy engineering was a retrograde step. The last thing you
notice about a paranthropine skull is the braincase. Yet it would be
a mistake to see the paranthropine adaptation as a bad call in
favour of brawn against brain. Paranthropines emerged around
2.5 million years ago, and lasted till a million years ago, long after
the disappearance of the gracile australopithecines. Ranging from
the north-east to the south of Africa, they were a successful variation on the two-legged primate theme.

As Bernard Wood points out, the finds at the Ethiopian site of
Gona indicate that the Oldowan tool industry also appeared 2.5
million years ago, remaining essentially unchanged for a million
years. One of the paranthropine groups occupied approximately
the same period, and also ended up much as it began. No hominid
species is known to coincide with the Oldowan in this way; indeed,
there are only a handful of fossil fragments older than two million
years that can be attributed to *Homo* at all. In the absence of strong
evidence to the contrary, Wood argues, we should consider
Paranthropus boisei as a possible author of the Oldowan.[37] His suggestion is in keeping with the sentiments behind the name of the
genus, which means 'man's equal'. Whether *Paranthropus* made
tools or not, it would be unwise to assume that the *Homo* lineage
contained a spark of human genius from its inception. At the
outset, the key difference between *Homo* and other hominids may

not have lain in brainpower, but in a more catholic appetite. The two lineages represent opposite responses to a phase of cooling climate, the same phenomenon that had previously led to the separation of apes and hominids. Once again, primates had to respond to the retreat of the woods and the advance of the grasses. *Paranthropus* invested heavily in extracting good-quality food from difficult plant sources, though it may also have eaten a certain amount of meat. *Homo* went further in its omnivorous tendencies, eating more meat and thus enjoying the benefits of a diet richer in energy. That left the hominids with a wider range of possible habitats, while the paranthropines remained confined to the tropics.

Omnivorous primates tend to have large brains, and it has been suggested that their cognitive capacities are enhanced in order to find a wide range of high-quality foods. Paranthropines were moderately omnivorous, and their brains were moderately larger than those of the gracile australopithecines. *Homo*'s more emphatic commitment to omnivory may have been responsible for both of the two spectacular expansions that distinguished it from other hominids; in geographical range, and in brain size.

Its record also expands – explodes, almost – just after two million years ago. Before that point, there is just a cryptic fossil image or two; a mandible at 2.3 million years, some teeth that could be as old, or older, both age and type indeterminate. After it, at around 1.9 million years, the record stages a major exhibition of remains from African sites. Meanwhile the first possible sightings beyond Africa are being registered, out as far as China. All of a sudden, it seems to be raining hominids.

Around two and a half million years ago, then, the paranthropines and the hominids split away from the australopithecines. At the same time, the first stone tools were made. The paranthropine lineage diversified, but these creatures changed little in comparison to the hominids. Within half a million years, *Homo* brains had increased markedly in size, reaching about 750cm^3. *Homo* evolution is not all onward and upward, though. The types extant in Africa at the 1.9 million year mark included species named *habilis* and *ergaster*. *Homo habilis* resembled australopithecines below the neck. It was not fully adapted to a life ranging across open country. *Homo ergaster*, however, was taller and classically human in its body type.

The same is true of its Asian counterpart, *Homo erectus*. The two species are broadly similar in form and brain size, but *erectus* has a larger face, and its skull is more heavily built. Both types probably evolved from a common ancestor, which expanded beyond Africa before two million years ago.

By a million and a half years ago, hominids had long since been settled across a great arc stretching from southern Africa to Java. Their brains continued to grow, reaching 800 or 900cm^3. That suggests an evolutionary dialectic, larger brains helping hominids cope with new environments, which in turn exert selective pressure for increased brain size.

This is still not a simple story of hominid progress, though, because the tools stayed the same. The artefacts from the Ethiopian site of Gona, announced in 1997, impressed their finders not just with their dates, the oldest yet, but with their sophistication. The archaeologists considered that the objects were as advanced in basic technique as tools made at the other end of the Oldowan phase, a million years later. But other artefacts from within this period are less complex, and had hitherto been regarded as a more primitive industry that preceded the Oldowan. To put it bluntly, they were made by smashing stones together, with little insight into how stone fractures. This is what Kanzi, a bonobo, has learned to do, with human encouragement. Born at the Yerkes Field Station in Atlanta, Georgia, Kanzi is the star pupil among the succession of apes who have been hothoused by human educators. Sue Savage-Rumbaugh and the other scientists who have worked with him are keen to encourage him to be the best he can be, but they have their doubts about whether apes could ever knap stone to Oldowan standards.

Kanzi's example also emphasizes the importance of human culture to ape achievement. The primatologists tried and failed to teach his adoptive mother, Matata, how to communicate using a special keyboard, the keys of which bore graphic signs representing words. Meanwhile, however, Kanzi began to pick up the vocabulary as he hung around. By the time he was six, he had learned 200 symbols. Whereas Kanzi had spent his entire life in an environment created by humans – he grew up in a fifty-acre wood – Matata had been caught in the wild.[38]

As the Gona team suggests, there may indeed have been a phase in which hominids smashed stones but did not knap them. Once they got the feel of stone fracture, the development of knapping techniques could have been a sudden leap forwards. The story that unrolls from the archaeological record suggests that once the knack of stone-knapping was grasped, the level of craft remained static for a million years. Over this span, many artefacts were made using simpler techniques. One explanation might be that different species made different grades of tool; another is that different tools were produced according to environmental circumstances. And sometimes the hominids may simply have lacked the necessary skill. The accomplished Gona tools are set off nicely by finds from Lokalalei, in the Lake Turkana region, laid down 100,000 or 200,000 years later. Mzalendo Kimunjia describes a rather sorry assemblage of badly struck flakes without sharp edges, and chips too small to use. It was as though they were trying to knap, rather than smash, but just hadn't got the hang of it.[39]

One of the attractions of this scenario is that it seems to prefigure what happened when the Acheulean emerged. Although there is something of an embryonic phase in which bifaces are moving towards the Acheulean form, but have not quite got there, the Acheulean is born abruptly rather than gradually assuming shape. As with the Oldowan, simpler forms continue to appear alongside the newer types. And, also like the Oldowan, the

Acheulean is notable for stasis rather than progress. There is a rapid climb to a higher level of complexity, and then a long sojourn on this plateau; or at most a crawl up the gentlest of inclines.

Yet the Acheulean industry has the aura of success, since it lasted a million years and spread so far. The secret of that success is enigmatic not just because of the elusive secret of the handaxes themselves, but because the Acheulean cannot be pinned to a new hominid species. *Homo ergaster* invented it, in Africa, long after the species appeared; *heidelbergensis* was still making it, in Europe, long after *ergaster* had disappeared.

Like the idea of progress, though, prehistoric success may not be what it seems. The idea that Man always has his eyes on the horizon, always wants to push back frontiers, appeals powerfully to the myths of Europeans and European Americans. It is easy to equate prehistoric movement with colonial expansion, and to imagine the most mobile hominids as the most advanced. According to Roy Larick and Russell L. Ciochon, though, the Acheulean industry may have been more a way of keeping its makers in Africa, rather than enabling them to range ever wider. They believe that early *Homo* was a scavenger, and adopted a simple strategy to make the most of the landscape it occupied. The oldest stone tools have been found near major topographical features such as lake margins or rock outcrops; these, Larick and Ciochon argue, were 'catchments' where resources were concentrated. Hominids lingered at these sites, which provided them with several essential resources – water, stone, food, shelter – in a single location.[40]

The disadvantage of one-stop hominid shopping was that as the land dried out, catchments became fewer and farther between. Hominids had to disperse in order to reach the locations that could support them. As they moved through Asia, however, they may have found it easier to scavenge or hunt among species not yet familiar with hominids, and may also have enjoyed the benefits of more temperate regions containing fewer parasites. Their rudimentary tools sufficed in these conditions. Meanwhile in Africa, after two million years ago, more sophisticated hominids developed 'territory scavenging', a more complex strategy which allowed them to make more efficient use of the resources in an area. It is characterized by visits to sites containing only one or two resources, and requires a high degree of local knowledge. By allowing populations to remain in Africa, territory scavenging would have increased competition between hominid groups. The strategy became common at the same time that the Acheulean industry was established.

A couple of hundred thousand years earlier, hominids in the Lake Turkana region seem to have taken a number of important steps, many of them literally. They suddenly began to make use of a wide range of habitats, and to impose standard patterns upon their tools.[41] It was at this point that the Oldowan industry began the development that culminated in the Acheulean. Perhaps handaxes are the hallmarks of a new way of using the land – or of rivalry over it.

Handaxes spread like a brush fire from the region where they first arose, appearing in what are now Ethiopia, Israel, Kenya and Cameroon 1.4 million years ago. In geological terms, that is instantaneously. Even if the earliest handaxes are a hundred thousand years or two older, as some estimates suggest, the Acheulean still reached its point of take-off pretty rapidly. This suggests that handaxes were the hallmark of populations which had developed ways of living much more efficient than previous hominid behaviour patterns.

With the announcement of the early dates for Asian hominids, there is now a straightforward explanation for the absence of Acheulean assemblages in eastern Asia: hominids in the region were descended from settlers who left Africa before the invention of handaxes. That may also account for the absence of early handaxes at the western end of the Eurasian land mass. There are none in European sites claimed to be a million years old or more. Nor are they found at the 780,000-year mark at Atapuerca. The Boxgrove collection, from half a million years ago, is old by European Acheulean standards; late by world ones. One possible explanation is that the first hominids in Europe did not come directly from Africa, but were descended from the eastern Asian population. Handaxes were the signature of a direct influx from Africa, occurring around half a million years ago.

Even once the Acheulean was established, other industries persisted. In southern England, many locations contain handaxes, while others yield assemblages of irregular tools that look more primitive. The Acheuleans' unpolished cousins are named after the East Anglian seaside town of Clacton, where the specimens used to define the type were found.

In less circumspect times, archaeologists used to assume that different types of artefact represented the work of different races. The Clactonian knappers were taken to be a different, and less sophisticated, group than the Acheulean people. Now it is recognized that the same hominids were responsible for both industries. A subtler account is needed to explain why sometimes they cared about form, and sometimes they did not.

8

The Boxgrove hominids were undoubtedly butchers. The handaxes lie by the bones, and the bones are scarred with the kinds of marks left by stone tools, not by animal teeth. It's a Palaeolithic smoking gun. There has even been a reconstruction featuring a modern butcher. John Mitchell, an Oxford research student, decided that somebody who cuts up meat for a living might make a better judge of a handaxe's mettle than somebody with a Ph.D. He recruited a local butcher, Peter Dawson, to try out a batch of handaxes made in the Boxgrove style, from Boxgrove flint, by Phil Harding. Dawson had no difficulty stripping the flesh from a roe deer carcass, and pronounced one of the bifaces to be 'a perfect butchery tool'.

In *Fairweather Eden*, a popular account of the Boxgrove Project by Michael Pitts and the Project's director Mark Roberts, the account of Dawson's handiwork is intercut with passages from academic commentaries on the mysteries of the handaxe.[42] One by one, it is implied, the scholars are felled by Dawson's slices: Richard Leakey and Roger Lewin, claiming that 'no one can think of a good use' for handaxes; Thomas Wynn, reflecting that the handaxe 'does not fit easily into our understanding of what tools are'; Eileen O'Brien and William Calvin with their killer Frisbees; and Iain Davidson's declaration that 'the "handaxe" was a by-product of stereotyped motor patterns producing flakes for use, with no decisions beyond the removal of the next flake'. All humbled by the tradesman.

Davidson may deserve the jibe for his lofty dismissal of Lower Palaeolithic hominid mental faculties. But the claim implicit in this set-piece, that the entire community of scholars has failed to spot the simple truth about the handaxe, is just as vain. Leakey and Lewin are guilty only of using conversational rather than precise phrasing. O'Brien and Calvin's ideas may be wild throws skywards, but they add to the gaiety of science, and in their improbability they honour the size of the problem they strain to solve. Thomas Wynn is thoughtful, circumspect, and right.

John Mitchell made a shrewd choice in recruiting a butcher. Within the strictly limited domain of butchery skills, somebody with a working lifetime's experience of cutting flesh can reduce the gap between unskilled modern humans and Acheulean hominids. Although Peter Dawson was used to steel knives with handles and straight edges, he quickly adapted to stone tools with curved edges and no handles. He rolled the edge through the carcass in an arcing motion, holding the tool between finger and thumb. This exercise, demonstrating that it is possible to cut up an animal of medium size without having to grip the axe tightly in the hand, raises the possibility that the continuous edge might be an attractive design from an economic point of view. The greater the length of sharp edge on a handaxe, the longer the tool could be used before having to be discarded.

It could also be used more efficiently if the shape of the biface allowed a substantial length of edge to make contact with the material being cut. According to Peter Jones, in the most persuasive of the functional explanations for handaxes, this consideration was the basis of the Acheulean industry. Jones points out that as the size of an edged tool increases, the weight increases faster than the length of the edge. Bifaces were a means of minimizing the extra weight. They served as a portable resource, suggests Jones, which could be carried around and resharpened in areas where it was difficult to obtain stone for knapping. Developed Oldowan tools were, in fact, Acheulean bifaces that had been reshaped and resharpened after use.

While the bifacial form, thinned on two sides, addressed the problem of increasing weight, the triangular shape optimized the tool's cutting efficiency. The closer a biface is to a circle in shape, the shorter is the length of edge that comes into contact with what it is cutting. Long thin tools with longer cutting edges required blade-making techniques that were not developed until much later; the teardrop shape was thus an effective compromise within the available technology between efficiency, weight and strength. All in all, the handaxe was 'a well-designed tool aimed at the problems of living and travelling through material-scarce areas'.[43]

It looks as though Jones may have worked out just why tools began to converge on a triangular form in the first place, as the Acheulean emerged from the Oldowan. But although Jones's model explains the beginnings of the handaxe, it does not account for its continuing story; its persistence across time and different environments, or its refinement beyond the call of efficiency. And set against the cogent suggestions about the efficiency of the form is one signal weakness. If the biface was a response to conditions in which resources were scarce, why have so many of them remained as bifaces instead of being knapped away into 'Developed Oldowan' tools? Although Jones claims that 'the average Acheulean handaxe is an optimum solution to functional and structural problems', he admits that handaxes beyond the average are not so neat. There are places, such as Olorgesailie and Isimila, where size seems to have been the main consideration in the

handaxe-makers' minds. At Olorgesailie, the one thing more striking than the size of the handaxes is their abundance. If hominids
needed to make larger bifaces because they were travelling further, as Jones suggests, then the system was not working at these
sites. Whether the handaxes were transported and discarded
without being used, or whether they were made and not transported, it was a spectacular waste.

Maybe the Acheuleans began to make handaxes in conditions
of scarcity and just carried on when conditions became less
demanding. But that requires two different forces to be invoked:
necessity to create the form, and inertia to maintain it. As well as
making the model more complicated, this would imply that force
of habit outweighed the extra costs of making handaxes. This
seems unlikely, since hominines did continue to make flake tools
as well as handaxes. If handaxes did not have some genuine
advantage, one would expect hominids to replace them with flake
tools where possible.

Peter Dawson did not have to pay for his tools either. Nor did
he shed much light on how well flakes compare to handaxes,
because he was so impressed with the handaxes that he refused to
try most of the flakes, 'despite appeals from the author', as
Mitchell plaintively noted in his report.[44] At Olduvai, Peter Jones
made the comparison, having learnt butchery techniques from
his local Wakamba colleagues. He found that flakes were best for
animals of moderate size, like goats, but handaxes sometimes
proved more suitable for larger beasts, such as cows or zebra.
Handaxes were less tiring to use on long jobs, lasting an hour or
two. They were also safer: he often cut his fingers when working
with small flakes, but rarely cut his hand with the handaxes. He
reasoned that the length of the edge worked both ways, cutting
more efficiently on one side, and dispersing the force delivered to
the palm of the hand on the other.[45]

Experiments like these have shown that flakes are adequate for
many tasks, better than handaxes for some, and they are cheaper.

The question is why hominids should have incurred the extra costs of making handaxes; and why, having gone to the trouble of making them, did they typically discard them so soon that the edges remained sharp enough to cut archaeologists' fingers half a million years later? Unswayed by rational economic considerations, handaxe users did not make the most of the continuous edges.

The bones and handaxes at Boxgrove were not the first evidence that hominids used handaxes for butchery, either. At Olorgesailie, the giant handaxes are sown in their thousands, and mingled with these dragons' teeth are bones with handaxe signatures. Nor is it surprising that meat featured prominently in the Boxgrove hominids' diet. With the climate similar to that of southern England today, they could hardly have survived through the winter on plant food.

Advocates of vegetarianism sometimes argue that the importance of meat in human prehistory has been exaggerated, and most anthropologists would probably agree that they have a point. 'Man the Hunter' was a figure from a period in which spectacular human fossil finds were adduced to an account of human nature told by intellectuals, as Donna Haraway argues in *Primate Visions*, awestruck by the capacity for violence demonstrated by humankind in the first half of the twentieth century. He reached his academic zenith in the late 1960s, with a conference of the same name. This was also the point when he made his most celebrated appearance in popular culture. In the opening sequence of Stanley Kubrick's film *2001: A Space Odyssey*, inspired by the ideology of Dartism-Ardreyism, an australopithecine smashes a skull to pieces with a bone. According to the killer ape vision of humanity, such was the defining moment of hominid evolution.

Man the Hunter was challenged by Woman the Gatherer, a feminist construction with a considerable quantity of evidence behind her. Anthropologists presented data indicating that foods obtained by women were the staples of contemporary hunter-gatherer diets. Women's foraging, mostly of plant foods but including a certain amount of animal material, such as small game or shellfish, might account for 80 per cent of a group's food intake. Men, by contrast, would only rarely return from a day's hunting with a kill. Latterly, since Western men learned to promote themselves by standing male braggadocio on its head, accounts of hunting have become an opportunity to mock male pretensions. Jared Diamond provides the definitive example in his book *The Rise and Fall of The Third Chimpanzee*, where he describes being out with a dozen New Guinean hunters, when they suddenly galvanized themselves into action as they spied a target. Diamond began to look for a tree to climb, to escape from whatever mighty beast they were stalking, but the prey turned out to be two baby wrens.

To make sense of foraging strategies, anthropologists have analysed hunting from a neo-Darwinian perspective. Hunting may not appear to be particularly beneficial to the group – but modern Darwinism is based on the insight that natural selection acts on individuals, not groups. Although it often produces effects that happen to be good for groups, it does so because the traits for which it selects are good for individuals. The usefulness of a behaviour is measured by how it contributes to an individual's reproductive success. If a male hunter were to return with a bag equivalent to what a plant forager collects, modest but adequate and reliable, he might be able to meet the needs of a mate and her offspring. Big game, however, offers different possibilities. Although it comes rarely, it comes in large packages. The meat cannot be stored, so it is shared out. This allows a hunter to show off his skills, and to favour particular individuals with meat.

Among chimpanzees, the other notable hunters in our neck of the primate woods, hunting is a means to sexual and political ends. A successful hunter may give meat to other males, as a tactic of his coalitionary politicking, or to females, in exchange for sex. At

Mahale, in Tanzania, primatologists observing an alpha male called Ntologi deduced that he assigned his gifts according to two rules. One was never to give meat to up and coming young males; the other was never to give meat to the beta male, the individual posing the greatest threat to the alpha male's status.[46] Sexual possibilities may also influence whether or not a chimpanzee group decides to hunt. Gombe chimpanzees prey mainly upon red colobus monkeys, inflicting a heavy toll, but they do not hunt at every opportunity. Observers identified several social factors that appeared to influence the decision whether or not to hunt. One was the size of the foraging party. Another was the number of adult and adolescent males in the group. A third was the presence in the group of females with the swollen genital areas that indicate oestrus; and of the three, this was the best predictor of a decision to hunt.

The females' influence might be indirect. By attracting males, they could cause large groups to gather around them, and these would be keener than small groups to go hunting. Large groups make more successful hunting parties than small ones. But individuals in large bands do not secure greater amounts of meat than those in small groups. That suggests a motive other than nutrition; and the observation that a male is more likely to give meat to a female if she is in oestrus gives a strong clue as to what the motive is.[47]

There is some evidence that meat has a value as sexual currency in humans: a study among the Ache people of Paraguay indicated that better hunters have more sexual partners.[48] Modern foragers are not living fossils, though. The few that remain are in marginal habitats, so the options open to them may be just a remnant of former glories. Many foragers, such as the Zu/'hoasi (also known as !Kung) of southern Africa, have been surrounded by farmers for centuries. Their way of life may represent the best they can do with what the farmers have not taken, rather than what they could do if everybody still lived like them.

One study of Zu/'hoasi diet found that a mere 6 per cent of the calories came from meat. Ten times as much came from plant protein; at least 11 per cent came from a single source, the mongongo nut.[49] But the proportion of meat in the diet does not necessarily indicate its importance. Imagine a region where winter lasts three months, during which time no plant food is available. The hunter-gatherers who live there only eat meat during the winter: animal food only accounts for a quarter of their annual food consumption, but their survival depends on it. The ability to make use of meat might have been equally critical in less clear-cut circumstances. Robert Foley and Phyllis Lee estimate that in the hominid lineage, meat consumption levels between 10 and 20 per cent of nutritional intake 'may be sufficient to have major evolutionary consequences'.[50]

Occasionally, vegetarians and their sympathizers push their case beyond its limits by suggesting that meat actually doesn't suit us. 'Humans are essentially vegetarian as a species,' states the writer Alexander Cockburn, an omnivore himself.[51] But although anthropologists who study the lives of modern hunter-gatherers might agree that foragers are largely vegetarian in practice, those concerned with our evolution would deny that humans are essentially so. In fact, they see eating meat as fundamental to the development of our species. The reason, they explain, is that members of the *Homo* genus have had big brains to support; and brains are expensive. Although a human brain accounts for only a fiftieth of the body's mass, it demands nearly a fifth of the body's energy consumption. This demanding organ is 4.6 times as large as it would be if the human body were built like an average mammal. It would tip the scales at less than 300g, whereas it typically weighs a kilogram more than that. Per kilo, the brain's metabolic rate is nine times higher than that of the body as a whole. It is responsible for just under a fifth of the body's energy use, representing a power consumption of about fifteen watts. (Those lightbulbs that appear over cartoon figures' heads are much too bright.)

One way a hominid could balance its energy budget would be to increase its metabolic rate. That, however, would oblige it to spend much more time feeding, searching and competing for food. Since humans have metabolic rates similar to those of mammals in general, it looks as though hominids did not follow that strategy. To work out how humans can afford their expensive brains, Leslie Aiello and Peter Wheeler conducted an audit of the human body.[52] The brain is not the only hungry organ. The liver uses over 20 per cent of the body's energy, the heart 12 per cent. Weight for weight, the gut is as expensive as the brain; the heart and kidneys are considerably costlier. Like managers or ministers, Aiello and Wheeler have difficulty locating an area where costs can be substantially cut. The brain depends on the liver to regulate the supply of sugar, on which it depends for energy. When the kidneys produce urine, the fluid upstream is dilute; water is reabsorbed via the network of tubules through which the urine passes. Smaller kidneys would have shorter tubules, and so would allow a more dilute fluid to leave the body. In hot open country, that could be a critical disadvantage. These are, after all, vital organs. Muscle is more dispensable, but it is much too cheap. To pay for the extra brain, 70 per cent of it – 19kg – would have to be shed.

The only place where the necessary savings could be achieved is the gut, and it does indeed appear to have been the focus of the hominid economy drive. While the human brain is huge by primate standards, the human heart, liver and kidneys are the sizes one would expect if a typical primate were built on a human scale. The one expensive organ that is conspicuously small in humans, relative to body size, is the gut. As in the kidney, the effect of shortening the coils is that the organ can absorb less of what passes through it. To compensate, the system requires a diet which provides more energy and is easy to digest. Among the candidates as fuel foods are bulbs, tubers, nuts and meat.

The bottom line of the 'expensive-tissue hypothesis' is that the price humans paid for a large brain was a reduced gut. Shrunk to 60 per cent of typical primate proportions, it saves about nine and a half watts.

9

Surveying the record of fossil skulls, Leslie Aiello and Peter Wheeler detected two phases of notably rapid expansion. One begins with the appearance of the *Homo* genus, somewhere between 2 and 2.5 million years ago. In the earliest *Homo* fossils, classified as *habilis* and *rudolfensis*, average brain size is 650cm^3, compared to 500cm^3 in robust australopithecines. The phase continues until the appearance of *Homo ergaster*, with an 850cm^3 brain, 1.8 million years ago. About half a million years ago, the second phase begins, with the emergence of the type known as archaic *Homo sapiens*.

Not everybody detects this pattern; others consider that brain size increased gradually, without much variation in rate. The sparseness of the fossil record leaves plenty of room for different readings. Fewer than thirty pieces of skull have been attributed to *Homo habilis* since it was christened in 1964, while Aiello and Wheeler's figure of 650cm^3 for the type is based on just eight specimens.[53] Nobody, however, has suggested that the brain went through a growth spurt in the 400,000 years immediately preceding the appearance of the handaxe, 1.4 million years ago. The beginnings of the Acheulean correspond neither to the emergence of a new species, nor to a *Sturm und Drang* phase in the expansion of the brain. As far as the record goes, handaxes appear in the middle of a plateau.

The record says next to nothing about what might have been going on inside the skull during this period, though. The brain might have been deepening its fissures, elaborating its convolutions, furrowing its brows. Changes like these rarely leave their marks on the mould of the skull. Rarer still is agreement about such traces. Two scientists, Ralph Holloway and Dean Falk, have been locked in disagreement about the relief map of the australopithecine braincase for years. At the other end of the scale, dramatic changes might have been achieved by reorganizing the fields and paths of the cortex without increasing the size or shape of the whole. Being cheaper, changes like these are more likely to have been made than gross ones.

In a commentary published with Aiello and Wheeler's account of the expensive-tissue hypothesis, Dean Falk endorsed the idea of looking for physical changes that permitted the expansion of the brain, rather than trying to identify what selective pressures actually drove that expansion. She referred to such changes as 'prime releasers', as distinct from 'prime movers', of brain evolution. Falk has proposed a prime releaser of her own. Her 'radiator hypothesis' argues that a network of cranial veins evolved to disperse heat from the australopithecine head, increasingly exposed to the sun as the early hominids adapted to open country. This adaptation coincidentally set in place an infrastructure capable of delivering enough oxygen to fuel a larger brain.

Despite Falk's preference for keeping her theoretical feet on the ground, evolutionary science always itches to know not just what happened, but why it happened. The oldest candidate as prime mover is tool use. It was Darwin who first suggested that bipedality freed the hands, and the mind followed. Until quite recently, it was generally felt that 'Man the Tool-maker' had used tools to make himself. Tool-making and language were taken as the essential characteristics of humanity, the one just as important as the other.

Tools must surely have helped shape the mind to some degree, but they could not have been the prime mover of brain expansion. All the way through the timeline, the fossil and archaeological records are out of synch. Australopithecine brain sizes remained generally static for a million years before the oldest known stone tools were made, about 2.5 million years ago. Over the subsequent million years, fossils of the *Homo* genus attained markedly larger brain sizes, but hominid technical capabilities do not appear to have changed substantially during this epoch.

If tools were not the prime mover, language might seem to be the obvious alternative candidate. It is stymied, however, by the lack of traces it leaves in prehistory. Many scientists hold fiercely to the view that it only emerged very late in the evolutionary day. Others believe it began to appear very early, but can say little about how it developed before the so-called 'cultural explosion' of the last 50,000 years.

To seize upon the faculties in which we differ most radically from other primates may, however, be the wrong way to go about the search for a prime mover. Over the past couple of decades, our understanding of the mind's evolution has been transformed by insights about processes common to a host of species. One school of thought has concentrated on diet. In 1979, Sue Parker and Kathleen Gibson presented evidence that omnivorous primates have larger brain sizes, relative to body size, than ones with specialized diets. Parker and Gibson argued that omnivores needed extra cognitive capacity to extract nourishment of high quality from a wide range of foodstuffs.[54] Around the same time, Tim Clutton-Brock and Paul Harvey gathered data which showed that fruit-eating species have bigger brains than leaf eaters.[55] Their interpretation was that fruit eaters range over larger territories, and need larger mental maps to keep the distribution of food sources in mind.

Meanwhile, a second body of scholarship has arisen from an insight far more profound in its implications than ideas about feeding. It is not a new idea, but its time took a while to come. After a trip to study the lemurs of Madagascar in the mid-1960s, Alison Jolly suggested that in adopting a characteristic primate social structure, those endearing prosimians had met the vital pre-condition for a dramatic increase in intelligence, although they had not taken up the opportunity. There are no monkeys or apes on Madagascar; there are no lemurs elsewhere, which suggests that lemurs cannot cope with competition from other primates. (Unfortunately, their declining fortunes in the face of their one primate competitor offers overwhelming evidence in support of this idea.) Lemur troops are similar to monkey troops; Jolly pro-posed that such associations were the matrix and the stimulus for primate intelligence. The 'rudiments of primate society preceded the growth of primate intelligence, made it possible, and deter-mined its nature'. As the monkey and ape lineages evolved, social integration and intelligence raised each other up in an 'ever-reinforcing spiral'.[56]

Ten years later, in 1976, the neuroscientist Nicholas Humphrey proposed that primate intellectual faculties have 'evolved as an adaptation to the complexities of social living'.[57] By that stage, his peers were ready to develop this idea further. Since then, and particularly in Britain, theorists of mental evolution have recog-nized that humans are social animals above all, as were their ancestors. The prime mover was life in groups. When it came to the mind, natural selection was other hominids.

10

We've seen the stage and now it's time to see some players. A party of hominids has made its way down the gully where the stream has taken the edge off the cliff. Passing beyond the cliff base, where the flints are shed from the chalk, they roam over the littoral plain, searching for game. Unusually, they surprise a solitary horse and bring it down with spears. The pack converges on the stricken animal, subdues it with spear thrusts, then finishes it off using cobbles and flaked stones.

At this point, the hominids' actions do not appear to alter greatly. There is no clear behavioural transition from the kill to the butchery: to put it more bluntly, the hominids begin to hack pieces off the beast before it is dead. It is in the interest of each to do so, because while these hominids co-operate in the kill, they compete in the butchery. Although the actions are much the same, the strategy behind them is transformed, from collective to individual ends.

Half a dozen of the hunters secure limbs or large blocks of the carcass; more than they can consume themselves. They command attention from the others, even while the rest of the field is carving out its modest portions. They move back from the remains, and knap handaxes. Some work from rough-outs; one starts with a fresh nodule of flint. Each then uses his tool to strip the flesh from his take. The party sets off for the high ground, carrying flesh and handaxes, but along the way they discard the latter, still sharp enough to draw blood.

Or, the party pauses at the base of the cliff, where each hominid finds a nodule of flint and knaps a handaxe from it. The group then sets off for the waterhole, where deer and rhinoceros frequently gather. They succeed in wounding a deer as its companions bolt, then deflesh it with their axes. Meat and antlers are taken away; bones and handaxes left behind.

These are the sorts of scenes typically portrayed in the paintings which illustrate popular books and museum displays about human prehistory. Instead of huntsmen on horseback, hunter-gatherers pursue horses, but the elements of composition are similar. Critics of the genre have identified borrowings from other dimensions of culture, too. They have questioned why hominids have so often been depicted in nuclear family groups, why representative figures are almost invariably male, and why it is always the male who is clutching the weapon so tightly that one might imagine it was part of his anatomy. Practices have altered as a result, and the gender roles in hominid portraits are now studiedly blurred. The days in which prehistorians could speak of Neanderthal 'housewives' are long gone.[58] So it would be reasonable to ask why, in the first scenario above, the alpha hunters are said to be male.

The trouble with paintings is that everything has to be filled in. The artist may have nothing to go on beyond the shape of the face, but is nevertheless obliged to make a guess about the hair. Writers are freer in their ability to proceed in the absence of information, but are all the more accountable for their choice of words. They also work from readings of other texts, and the scenarios here are readings of texts about Boxgrove, shaped to illustrate my vision of what handaxes really mean. The use of the male pronoun is not an assumption, or a political stance, but a proposition.

What archaeologists do is also reading, according to contemporary critical usage, and extremely close reading at that. At Boxgrove, the analysis of the horse butchery site also entailed a translation. This site epitomizes the true value of Boxgrove; not as the spot where a chewed shinbone and two teeth give Britain a claim to the premier league of first European hominid fossils, but as a cabinet of maps that indicate hominid behaviour in fossils and stone fragments. At the horse butchery site, pieces of horse bone, including parts of a shoulder-blade and a femur, lay close to each other. Around them was a litter of tiny flint chips. The positions of these objects were recorded before removal. Then they were all laid out again in their original pattern, on an old carpet in one of the barns. As the pieces of flint were compared, to see which belonged together, six or seven concentrations of debris became apparent. One collection was refitted to form a flint nodule, with a handaxe-shaped hole at the kernel. The scene was read as the 'encounter' kill of a single animal, in which the hominids had happened upon their prey without having prepared tools beforehand. The concentrations of flint chips were the *débitage* left from the knapping of six or seven handaxes on the spot.

Overall, Mark Roberts reads the record of butchery at Boxgrove as a testament to the sophistication of Lower Palaeolithic hominids, and as a refutation of the prevailing tendency to discount their capacities. Most importantly, it provides strong evidence for hunting, as opposed to scavenging. Vegetarians and feminists were not the only adversaries Man the Hunter faced in the 1970s and 1980s. His image waned and flickered all the more in the face of trenchant criticisms from Lewis Binford, who in the 1960s had set out the framework for a 'new archaeology'. This attempt to make the discipline a science rather than a form of history was eventually judged a disappointment, but it did succeed in getting archaeologists used to thinking in evolutionary and ecological terms.

While studying an Inuit group in Alaska, Binford observed the behaviour of wolves, and the remains they left of their kills. He noticed similarities between these assemblages and those in Africa which had been adduced as evidence of hominid hunting. Starting with these field observations, he developed a critique of claims about hominid hunting and the creation of 'home bases', a practice argued to be part of the hunting way of life. Hominids ate meat, he agreed, but usually it was already dead when they found it.[59]

The bias in interpretation, he argued, had its foundation in a nineteenth-century archaeological tradition whose primary concern had been ancient monuments. When archaeologists investigated profoundly ancient remains, far older than any civilization, they imposed the same framework that they used to investigate classical ruins. An assemblage was treated as a habitation, and everything within it was assumed to be there by the hand of Man – whether or not this was consistent with the principle that, in a scientific analysis, the most parsimonious explanation should be preferred. A circular hole in a horse's shoulder-blade might have been made by a spear or a wolf's tooth. If wolves are known to have inhabited the area during the period in which the bone was punctured, but no spears of the same vintage have been found, then parsimony dictates that a wolf's bite is a better account of the evidence.

It so happens that there was a circular hole in the shoulder-blade found at the horse butchery site. Although the fossil record shows that wolves, hyenas and lions shared the area with the hominids, the Boxgrove team interpreted the perforation as the work of a spear. Ironically, it is thanks to Binford that claims like these carry more weight nowadays, because his critique has raised the standard of evidence required. The crucial evidence discovered by the Boxgrove investigators is that where bones bear marks

of both stone artefacts and carnivore teeth, the toothmarks are on
top of the butchery scratches. In other words, the hominids got to
the carcass first, suggesting that they were the killers and the carnivores were the scavengers. The pickings were slim, judging by
the lack of toothmarks on bones with cutmarks, which attests to
the efficiency of the butchery. Then there are the three butchered
rhinoceroses, which would not have been vulnerable to any of
the carnivores known to have been present. This looks like hunting, not opportunistic scavenging. Hominids would have had to
outrun the lions and hyenas and, as Mark Roberts has pointed out,
keeping carnivores away from a carcass is a lot easier than dis-
lodging them once they have taken control of it. For a
demonstration in miniature, try taking a bone out of a cat's mouth,
and listen to the menacing growl of the wild.

When there's a prime suspect in a killing, the investigation
demands that a weapon be identified. It is hard to imagine how
hominids on flat ground could have killed a horse without the use
of throwing spears.

And such weapons have now been unearthed, albeit some distance from the scene, in the middle of northern Germany, and
dating from a somewhat later period. The three spears found at
the Schöningen open-cast brown coal mine are arresting proof of
Lower Palaeolithic hominid capabilities. They were carved,
400,000 years ago, from the trunks of spruce trees, sharpened at
the base where the wood is hardest. Two metres long, they were
thickest nearer the point and tapered back towards the other end,
like a modern javelin. Shorter lengths of wood were found with
grooves in them, as if to hold stone blades. Hartmut Thieme,
who described the find in the journal *Nature*, noted that this
would make them the oldest known tools made from more than
one component.[60]

The spears demonstrate an ability to assess the quality of raw
material other than stone, a grasp of ballistics, and a facility for
ergonomic design. Capacities like these add up to a formidable

technical intelligence, to use a term favoured by the archaeologist Steven Mithen. They set the conservative and sub-optimal design of the handaxe in even starker relief. When it came to tools, these hominids were not stupid. They could innovate, and improve the efficiency of their designs. According to Thieme, they practised 'systematic hunting, involving foresight, planning and the use of appropriate technology'. So why did they stick so rigidly to the handaxe?

Clive Gamble's negative vision of Boxgrove hominid society was like John Lennon's song stood on its head: imagine no conversations, no burials, no contact with strangers. Individuals made tools by themselves, for themselves. Mark Roberts's riposte lists the busy and diverse activities he believes the Boxgrove hominids pursued – hunting, carrying meat and skins back to camp, making wooden tools, taking good care of soft antler hammers – and argues that these must have needed language to knit them together. In particular, the degree of planning and co-operation demanded by large game hunting on open terrain would have demanded the use of speech.

At the horse butchery site and other encounter kills, where handaxes were knapped on the spot, 'the carcass would have had to have been secured by part of the group, whilst the other part went off to the cliff-line to collect raw material'. Roberts implies a co-operative spirit and a capacity for organization, but the effect could also have been achieved by the invisible hand of self-interest. If one or more individuals had left the carcass, there would have been all the more incentive for the others to remain. According to the calculus of inclusive fitness, it would then have been in the interests of the individuals surrounding the carcass for its resources to go to their fellows, with whom they shared genes, rather than to intruding hyenas. A hominid would not have needed to act differently, however, if it was acting solely on its own behalf. Whereas hunting tactics would have demanded co-ordination, defensive procedures – threat displays, noise, throwing

stones – could have been the same whether a hominid was on its
own or with its fellows. Keeping competitors at bay, whether they
had two legs or four, would have been much easier than displac-
ing them. Possession is nine-tenths of the law.

Nor are there any grounds for supposing that the hominids
were subject to any higher forms of law. Moral codes are embed-
ded in spiritual beliefs and abstract symbols. Clive Gamble's
vision of a social limbo is uncontestable in this respect. How much
brotherhood there was among the hominids is a matter of conjec-
ture, but there is no reason to suppose they ever imagined a
heaven or a hell. Nothing remotely reminiscent of behaviour
guided by a spiritual sense occurs in the record until the Middle
Palaeolithic, starting around 80,000 years ago, when complete
skeletons become much more common. These have been widely
seen as evidence that the Neanderthals buried their dead; as have
signs suggesting deliberate interment, such as a characteristic posture with legs flexed. The pendulum of archaeological opinion
has swung away from faith in Neanderthal spirituality, however.
When pollen was found associated with Neanderthal remains at
Shanidar, in Iraq, it was hailed as the remains of flowers placed
lovingly on the body. It has since been reinterpreted as a coincidental trace of the local flora, and other relics found with
Neanderthal fossils also tend to be regarded as accidental associations. Some archaeologists are prepared to accept that the
evidence for burial is good, but deny that it meant anything more
to the Neanderthals than a convenient way to dispose of decaying
corpses. Unequivocal relics of art and adornment do not appear
until the last 50,000 years.

Without such a symbolic order, it is futile to imagine that at the
horse butchery site, any hominids who left the carcass were going
to make tools for the benefit of the group, or that the others would
save a place for them till they got back. Walking away from a carcass would have been an expensive exercise. And that, I suggest,
is just why they did it. They were deliberately incurring costs, in

order to demonstrate their Darwinian fitness by showing that they could afford to take chances. This was an instance of what is known in modern evolutionary theory as the Handicap Principle.

The balance of interests usually turns the arrow of mate choice towards the male. When females have to invest a great deal in reproduction, while males so often get away with so little, it is in the interest of the female to pay closer attention to the quality of a potential mate. Females therefore choose males, broadly speaking; as a result, the traits on which they base their choices become more widespread, and may be magnified as well. This is a positive feedback process, and its momentum can become enormous. Imagine a time when peacocks had short tails. A peahen developed a preference for peacocks with slightly longer tails; her offspring thrived because peacocks with longer tails happened to be healthier than their fellows. The preference for longer tails thrived with them, and longer-tailed peacocks came to enjoy a double advantage. Not only were they healthier, they were increasingly preferred as mates. The latter advantage grew in importance, and the tails grew more splendid with it. Ronald Fisher, the biologist who published the first description of the process in 1930, called it 'runaway' sexual selection.

The peacock's tail is now as costly as it is spectacular. According to Amotz and Avishag Zahavi, whose theoretical speciality is brilliantly counter-intuitive evolutionary proposals, its cost is in fact its point. The Zahavis began by attempting to explain the peacock's tail, and they have developed their understanding of what they termed the Handicap Principle in their field studies of birds known as babblers.[61] When a babbler sees a raptor, or bird of prey, it gives a loud 'bark' call, which may be audible more than half a mile away. The usual explanation for behaviour like this is that it is a warning to the rest of the group. As a response to a threat, though, the babbler's bark seems downright perverse. Babblers are drab birds which can easily evade raptors by taking refuge in the thickets they frequent. They do not need to make loud calls to attract each other's attention, and their routine calls are soft trills. Oddest of all, they emit the 'barks' when a raptor is a long way off and would not have spotted them. If they just stayed put and kept quiet, the raptor might well overlook them altogether.

Eventually, the penny dropped. The Zahavis realized that the babblers' signals might be directed not at each other, but at the raptors. This was in the interests of both parties. If the babblers inform a raptor that they have seen it, the raptor will avoid starting an attack that is bound to be fruitless, while the babblers will be able to continue their normal business instead of taking cover.

For the arrangement to work, the signal has to be genuine. In other words, a babbler should not bark unless it really has spotted a raptor. It demonstrates its good faith by barking from the tops of trees, rather than just remaining in cover and barking from time to time on the off-chance that a raptor is in the area. When it climbs to the top of the tree, it may expose itself to raptors it has not seen from lower down, and while there, it is prevented from doing other things. Barking therefore carries a cost. So does stotting, the vertical four-footed jumps that a gazelle makes when it is approached by a wolf. This gesture demonstrates physical fitness, and signals that the gazelle can afford to give the performance without compromising its ability to escape if necessary. As with the babbler's bark, the price is the guarantee. The Handicap Principle rests on the axiom that signals have to be costly to be reliable.

When Amotz Zahavi first published the Principle in 1975, it met with a sceptical reception. John Maynard Smith and others used mathematical models to show that it would not work. But the scholarly community was intrigued as well as provoked, and some theorists stuck with the Principle. In the process they subjected it to a sort of editorial process, exploring ambiguities in its early formulation. Within a few years, different mathematical models showed that it could work after all.

Among those who shook it down was Alan Grafen, who identified four possible senses in which the Principle could be understood.[62] The first was that a handicap is a sort of test, which the bearer passes by surviving despite the burden. As the handicap is obligatory, all individuals have to bear its costs.

The second, which John Maynard Smith called a revealing handicap, conceives it as a special task rather than a permanent imposition. Casting an eye over a group of men who are standing around chatting may give some idea of their relative fitness, but entering them into a kilometre race will really show their mettle. The costs are not compulsory, but those who pay less, or decline to pay at all, will be judged inferior to those who pay more.

There are also 'condition-dependent' handicaps, which indicate the health and fitness of their bearers. Shiny and trim plumage indicates that a bird is in good condition, for example. Only individuals of high quality can display these handicaps; so only they pay the costs.

Finally, a handicap may be a matter of choice. In such cases, individuals decide how much of a handicap to take on, according to what they feel they can manage, and how much they are likely to gain as a result. This class of handicap offers the greatest flexibility, since individuals of whatever quality can vary their burdens. Grafen therefore sees 'strategic-choice' handicaps as a more comprehensive way of fulfilling the intentions behind condition-dependent handicaps. Condition dependence works because a low-quality male cannot increase his handicap level, whereas strategic choice works because it wouldn't pay him to do so. Carl Bergstrom of Stanford University observed that strategic choice is the theme of greatest interest in current thinking on the Handicap Principle.[63] It is also one which hominids are particularly well suited to practise, with their highly developed cognitive powers and their behavioural flexibility. They would have had considerable scope to make choices about costs when making handaxes.

From his calculations, Grafen concluded that the key to a stable signalling system was quality. The kind of signal given had to reflect the quality of the individual giving it, and signals had to be more expensive for signallers of poorer quality. The maths confirmed what we know so well from life in our own species. A wealthy man can afford a more expensive suit than a poor man, and it costs a poor man a greater proportion of his resources to acquire a suit at all. Both the rich man and the poor man will strive to demonstrate that they can consume conspicuously, as Thorstein Veblen recognized in his *Theory of the Leisure Class*, published in 1899, anticipating *The Handicap Principle* by nearly a century.[64]

The Zahavis have now applied the Handicap Principle to a whole ark's worth of species, including tigers, toads, zebras, skylarks, seals, sheep, dogs, ants and yeast. Armed with this menagerie, they have proposed the replacement of the two major varieties of selection, natural and sexual, that have stood since Darwin first defined them. Instead, the Zahavis distinguish between utilitarian selection, which covers everything except signalling, and signal selection. Utilitarian selection reduces costs and promotes efficiency; signal selection raises costs and promotes waste. Human language aside, they argue, all signals rely on cost.

A widespread form of signalling to which the Handicap Principle has been applied is the crying of the very young, including human babies. Babbler fledglings often cry loudest when their parents are close at hand and are well aware of their whereabouts. This is blackmail, say the Zahavis. The calls are directed at any predators in earshot. 'The fledglings say, as it were, "Cat, cat, come and get me! I am here and I don't care who knows it until my parents feed me".' As for cats, the kitten up a tree is not miaowing to tell its mother where it is. 'We think the kitten is trying to blackmail its mother into continuing to nurse it, to prevent the kitten from wandering off to dangerous places.' Sometimes the juvenile's luck fails it, and a predator carries it off.

But this is a necessary element of the system, for if the risk were not real, the cost would be small and the signal would not be honest.

Many of those who aren't yet ready to go all the way with the Zahavis might nevertheless agree that the Handicap Principle seems particularly well suited to explaining processes of sexual selection, which frequently involve the growth or construction of bizarre ornaments such as the peacock's tail. A healthy individual that carries a visibly costly burden is demonstrating its fitness through its handicap. The exhibitionist marathon runner who completes the distance wearing a waiter's uniform and carrying a tray of drinks is signalling that he is fitter than the runner in singlet and shorts who crosses the line at the same time. He is also the one who gets the attention.

In many cases, such as that of the peacock, sexual selection has forced a uniform burden upon all males. Among primates, displays may be more varied in strength and more diverse in character. Geoffrey Miller considers that hominids' complex social systems would have been peculiarly conducive to the sexual selection of mental capacities. He goes so far as to argue that the prime mover of brain expansion was courtship. Our brains evolved as courtship devices; the courtship displays they control, such as conversation and song, are adaptive but not innately specified. Devices used in sexual selection do not have to be coded in the genes.[65]

11

The handaxe makes partial sense as a tool for butchery or general purposes. As a Zahavian device, it makes perfect sense. A hominid who can knap a handaxe is demonstrating a number of critical capabilities. Handaxes demand the ability to conceive and implement a plan, despite difficulties that may arise. The first requirement is the ability to locate raw material, which in turn requires spatial skills and an understanding of patterns in the physical environment; the process of knapping demands the co-ordination of eye and hand, as well as physical strength. A handaxe is a measure of strength, skill and character.

It is also a litmus test of social fitness; of the ability to assess social relations, predict the behaviour of others, and assert one's interests within the group. These skills are at a premium if a hominid is to leave the horse carcass, go back to the cliff, pick out a flint nodule, return, knap a handaxe, and still walk away with a handsome quantity of meat. He has to be able to assess whether any of the others are going to make a bid to exclude him when he comes back, whether more than one individual may form an alliance against him for this purpose, and what any challengers' prospects of success might be.

To offset the disadvantage at which he is about to place himself, he needs to deter others from taking advantage of it. Their own social faculties may assist him, since they will have good memories for competitive form, and will recall previous occasions on which he has emerged from contests the victor. But all individuals' powers ebb at some point, and he will need to project the impression that he remains fighting fit. He may do so by making overt threats, or he may be able to rely on the very fact that he is prepared to walk away once again. Whatever tactics he employs, they will draw upon a significantly more sophisticated communicative repertoire than that available to chimpanzees today. Without this enhanced battery of skills, he would be unable to impress his image upon his companions' minds strongly enough to cover him in his absence. 'The alpha male's authority is enforced only by his presence,' the palaeoanthropologist Owen Lovejoy observed. 'If he goes down to the river for a drink, he loses it.'[66]

While the hominid is knapping his handaxe, close to the carcass, he has more opportunity to assert his interests – not least because of the large number of sharp stone flakes he has to hand. Precisely because of his handicap, he will command attention. Potential challengers will be looking for opportunities to seize any meat he is guarding; while he may attempt to secure more of the kill by means of threats and sallies towards the others. Keeping an eye on him, his companions will observe with interest the degree of skill he brings to the making of his handaxe.

Just looking at a knapper costs very little; which meets an objection to the idea of costly signalling made by Marian Stamp Dawkins and Tim Guilford in a paper entitled 'The Corruption of Honest Signalling'.[67] They argued that among animals, costly signals are also typically costly to receive. Female sage grouse, for example, prefer males who attend regularly at the courtship display rituals known as leks. To identify the regulars, the females have to incur the cost of regular attendance themselves. Receiving costly signals may not only be time-consuming, but increase the receiver's vulnerability to predators. Dawkins and Guilford argued that the prevalence of costly signalling is limited by expenses like these, and systems based on expensive signals will tend to develop only in the unusual cases where the signals are cheap to receive. Handaxe-making during butchery strikes just the right balance, demanding lengthy concentration from the maker, but only intermittent glances from the observers, who would be keeping their eyes on each other anyway.

Although the hominids are engaged in butchering a large animal, not building bowers, this event is the occasion for display with a significant degree of standardized patterning. That may seem like a ponderous way of saying it has a ritual dimension. If these were great crested grebes performing their staccato mirror routines on a pond, the term 'ritual' would spring to mind without danger of confusion. Since they are hominids, though, it's important to distinguish their patterned displays from those of their descendants, which are structured by symbols.

The details of the scene may not be played out exactly like this, but whatever the sequence of events, the Handicap Principle remains in operation. The hominid would incur extra costs if he carried a nodule of flint around with him in case prey was encountered, or if he made sure of his portion of carcass by carrying it to and from the cliff base when he left the kill site.

He might also have paid a portion of the costs in advance. At a workshop held as part of Steven Mithen's Reading University MA course in Cognitive Evolution, the professional flint-knapper John Lord illustrated how that might work. An expert in his craft – tact suggests he and Phil Harding be described as the two best knappers in the country – Lord knapped his way through a block of flint at a leisurely pace while he talked the class through the procedure. I asked him how long it would take him to make a handaxe of reasonable quality, working at his own rate. He reckoned he would have the job done in twenty minutes or so. On occasions when he had attempted speed trials, he had got it down to about twelve minutes.

That sounded like a long time when butchery was in progress, but then he offered to knap a quick handaxe on the spot. Starting from a small hunk of flint, rather than a boulder, he completed a decent biface in six or seven minutes. There was an invisible time cost, however, in the years of practice that had taught him to work so efficiently. The principle of paying for experience is illustrated by the tale of the tradesman who tells a householder that a repair job will cost £100. The customer agrees, whereupon the tradesman drives home a single nail and requests his fee. When the householder objects that the work had only taken a moment, the tradesman replies that he has charged £1 for hammering in the nail, and £99 for knowing exactly where to place it. Knowledge like that takes years to accumulate.

While young Acheulean handaxe-makers were in the course of perfecting their craft, they would been at a disadvantage compared to older, experienced knappers. If experience were the decisive factor, it could have led females to make bad choices. As the geneticist Steve Jones has argued, older fathers are more likely than younger ones to produce gametes containing mutations, as the copying fidelity of their genes deteriorates.[68] By the nature of mutations, most of them will be harmful. The handaxe handicap measures more than knapping ability, however. It points out individuals who can take care of themselves as well as making handaxes. Older knappers would only enjoy an advantage as long as they remained fast, tough and fit enough to hold their own. This would tend to exclude males of a dangerous age, genetically speaking. Nowadays other mechanisms produce the same effect. 'As far as the gene pool goes,' John Updike remarks in *Roger's Version*, 'we deliver our mail much earlier in the day than we like to think.'

Though the time costs could have been distributed in different ways, they could not have been avoided. Over the long term, many biface knappers would also pay a price in injuries. The risk of deep cuts would be much the same whether simple flake tools or bifaces were being knapped, since both types would require the kind of stone-breaking in which large wedges are struck off. Handaxe knappers would have been additionally vulnerable, however, because of the extra number of strikes required to complete a handaxe. The more blows are struck, the greater would have been the number of tiny splinters flying off on unpredictable trajectories, some of which would have terminated in a knapper's eye. Most would have done little harm, but a few would have caused lasting damage. In modern industrialized societies, institutions live in fear of the costs of accidents like these, which is why everybody in the room at Reading University was obliged to wear safety goggles while John Lord demonstrated his knapping skills. The costs of legal liability in modern societies pale beside the costs of such injuries in foraging societies. Blindness, partial or total, would be a catastrophe for a hunter-gatherer.

There again, evolutionary psychologists like to argue that males are more disposed to take risks than females. And when they do, they arouse the hackles of critics who suspect that sociobiology is an attempt to depict a temporary and contingent state of gender relations as an eternal verity. So far, this account has provided plenty of images to stoke such suspicions. Visualized, it looks like a return to Ardrey's killer apes, with co-operation only in slaughter, followed by a struggle of each against each as the carcass is torn apart. Dart and Ardrey saw the mark of Cain in the hominid fossil record, and of course Cain was male.

Jane Goodall, the doyenne of primatological field studies, saw real killer apes in the form of the chimpanzees of Gombe, in Tanzania. My sketch of the horse butchery site is based on her account of chimpanzee hunting, where the struggle for meat does become a war of all against all. Competition for portions of the kill may be intense and aggressive – though it pales by comparison with the violence wreaked upon the prey itself. 'Chimpanzees kill their prey by (a) biting into the head or neck,' she begins, '(b) flailing the body so that the head is smashed against branches, rocks, or the ground, (c) disembowelling it, or (d) simply holding it and tearing off pieces of flesh (or entire limbs) until it dies.'[69] Death may come instantly, or take three-quarters of an hour. The chimpanzees don't wait for it to arrive before they start eating. Goodall goes into relentless detail: the drinking of blood, the biting off of genitals, and the final flourish, 'The three large bush-pig young took between eleven and twenty-three minutes to die as they were slowly torn apart; the largest gave his final scream when Humphrey tore out his heart.' She calls these episodes 'somewhat gruesome'. Readers made of less stern stuff may wonder what would count as truly gruesome.

Appearances are deceptive, though. If you are susceptible to resemblances, a troop of monkeys is a disheartening sight. They have hands like ours and the piercing shaft of intelligence in their gaze. Their curiosity, their busyness and their speed is like a caricature of human activity; their grasping and pushing, the constant animation of self-interest, also reflects upon us. This, it seems, is the primordial primate truth. But because monkey life happens so fast, and many of the signals that guide it go over our heads, we can find it hard to see past the capering, cuffs and grabs. It takes scientific observation to appreciate the subtleties of monkey society, and to confirm that the intelligence in monkey eyes is not an illusion. Diagrams, not images, are the key to understanding. At Boxgrove, the archaeologists mapping the horse butchery site produced intricate diagrams in which tiny fingernails of flint were joined by criss-crossing lines, showing how the pieces had fitted together in the complete rock. Unfortunately it is not possible to draw a diagram showing the relationships and interactions between the hominids at the site, so imagination guided by reason will just have to do.

In circumstances like these, reason requires the selection of elements of the scenario on the balance of likelihood. That includes the casting of the sex roles. There is no unavoidable law of nature which ordains that males must be the most effective actors at scenes of butchery, nor that females must be the sole agents of sexual selection. Females may compete with each other for males of quality, and they may have an interest in retaining male partners to provide for their offspring. Evolutionary psychologists have discerned the landmarks of sexual selection upon the human female body; in breasts, buttocks, and the ratio of waist size to hip. As in sexually reproducing species as a whole, though, the difference in reproductive costs would seem to leave human males under more sexual selection pressure than females.

The present pattern of differences in form between the sexes in humans is consistent with such reasoning. J. Michael Plavcan, who has compared the data for sex differences among hominid species, has used a ratio of male to female mass among modern humans of 1.2: in other words, men are about 20 per cent heavier than women.[70] At the outset of the hominid lineage, differences in body mass between early australopithecine males and females seem to have been considerably greater, though Henry M. McHenry had to work carefully with ambiguous data to arrive at his ratio of 1.5. In between, the map is blank until the Neanderthals. With data points from the middle of the hominid timeline conspicuous by their absence, the graph is a straight line running downwards from the first hominids to the current ones. On that slope, the difference in size between the male and female hominids of Boxgrove would have been smaller than that between male and female australopithecines in Africa, but greater than the difference between modern men and women.

How much that says about competition between males is open to question, however. It might just mean that females got bigger, perhaps in response to increasing reproductive loads. Camilla Power and Leslie Aiello suggest that larger females would be able to travel long distances more efficiently, which would assist them in obtaining the higher quality diet they would need to support offspring with larger brains, and make it easier to carry their offspring. Power and Aiello also note that an increase in size might have made it easier to cope with a climate that was becoming hotter; and lactation might have become more efficient if mothers increased in size relative to their infants. With further increases in brain size, however, females would have needed to make use of male energy to meet the increasing energy costs of reproduction.[71]

The question is, what would be in it for the males? According to a theory put forward by Owen Lovejoy in 1981, the answer was that they would know who their offspring were. Lovejoy argued that although the hominids were distinguished first of all by their upright stance, the development that made them what they truly were was a sexual division of labour. Early hominid males did not hunt, but they gathered; females, hobbled by offspring, stayed in home bases, where their rates of reproduction increased. The foundation of this social order was monogamy, cemented by pair-bonding.[72]

And what a fine Aunt Sally the theory makes. It appeared too late in the day to get away with its transparent mapping of an idealized contemporary order on to savannah primates. Some of its weaknesses are almost as obvious. In her 'Reply to Mr Lovejoy', Dean Falk tartly pointed out that, if the females were so restricted in mobility, they and their offspring would be easy pickings for predators while their males were out gathering.[73]

Other events of major reproductive interest to the males might have taken place in their absence. While it echoed the family values espoused by conservatives, Lovejoy's theory was as politically utopian as those who thought that monogamy was the soft underbelly of patriarchy. Counter-cultural sexual politics nurtured the belief that jealousy and ties of blood were no match for ideals and good intentions. Lovejoy recognized the importance of relatedness, but imagined that all an australopithecine family needed was pair-bonding. As with attempts at egalitarian communal living, the idea was shaky at the time, and has become even weaker in the light of experience. Scholarly eyebrows have been raised by a series of studies showing how often fathers are not who they appear to be, among swallows, chimpanzees and humans.

The chimpanzee findings are the most surprising, because they show that despite male territorial defence, female chimpanzees are able to mate with males from other groups. The females of Taï, in the Ivory Coast, were found to enjoy such extensive freedom of choice that outsiders fathered half their offspring.[74] An australopithecine's confidence in his paternity would have been severely misplaced. In any case, it is hard to see where such confidence could have come from in the first place – or why the male should have stuck to his side of the arrangement, when extra-pair copulations, to use the technical term, might have cost him very little. Without guarantees, the way of life would have been wide open to cheating, which would have caused it to collapse. The weakness of unguaranteed bonds is illustrated by the observation of temporary associations, known as consortships, at Taï. Only two of the six male consorts fathered offspring, a considerably lower rate than that of the outsiders.

There is no reason to suppose that any basis for guarantees would have arisen by the time of the Boxgrove hominines. Robert Foley and Phyllis Lee have concluded from computer simulations that until the period during which the Acheulean ended, hominine females did not rely upon male provisioning to meet their increasing reproductive costs; which means that they were not monogamous.[75] Some food sharing would probably have taken place between related adults, and sexual partners. Jane Goodall and other primatologists have observed that a male chimpanzee in possession of meat will sometimes find that a female chimpanzee presents her rump to him. After coitus has taken place, the female receives some of the meat. It seems plausible that hominids would have conducted similar exchanges, their greater social sophistication encouraging them to do so more frequently and elaborately. But the safest assumption is that females could not rely upon such sources. They would have had to get the bulk of their food for themselves, and for their young.

If there was any division of labour, then, it would not have been at the heart of the hominids' behavioural repertoire. Hunting may have been a predominantly male activity, as it is among chimpanzees, but female hominids with large-brained off-spring would probably have had more need of meat than female chimps, especially in regions where the winters were cold. It is reasonable to suppose that they would regularly have been in at kills, just as female chimps sometimes are, except when encumbered with infants. In that condition, however, their interest in assessing possible mates would have been at its lowest ebb. They would also have been of little interest to males, since they would have been feeding their infants with milk, the production of which tends to suppress fertility. The absence of these females from episodes of butchery would not therefore diminish the potential of such events as opportunities for sexual display and scrutiny.

Females taking part in butchery or hunting would have used stone tools, otherwise they would have had no chance at all in competition with the larger and stronger males. That would have enhanced the usefulness of handaxes as a criterion for their choice of mates. The greater one's own accomplishment in a complex skill, the greater one's appreciation of others' accomplishments is likely to be. The capacities necessary for competence in knapping stone would have been enhanced in females through what Darwin called the 'Law of Equal Inheritance'. If females choose mates with a particular trait, their daughters as well as their sons are likely to inherit the trait. Darwin realized that these traits could be cognitive ones, and that his 'Law' could have reduced sexual inequality in humans. 'It is, indeed, fortunate that the law of the equal transmission of characters to both sexes prevails with mammals; otherwise,' he wrote in *The Descent of Man*, 'it is probable that man would have become as superior in mental endowment to woman, as the peacock is in ornamental plumage to the peahen.'[76]

Deciding whether the scenario should include females knapping handaxes, rather than simple flake tools, is a nicer judgement. One reason to suppose that females might have done so is the influence of example. Females constantly saw males making handaxes; they did likewise, and the making of handaxes became a general habit. It was a culture, but not a symbolic one.

Among modern humans, imitation of this kind is taken for granted, but scientists nowadays are cautious about attributing it to other living primates. When the habit of washing sweet potatoes in the sea was observed to spread among a group of Japanese macaque monkeys, it was hailed as a demonstration of culture in another species. The monkeys were taken to have learned from each other by imitation, a concept which was not closely scrutinized at the time. Subsequent study of primate learning has resulted in an unpacking of the term, and a deflation of claims for the complexity of learning in other species. Imitation is now considered to be the preserve of apes and humans: apes can ape, but monkeys can't. The Japanese monkeys are thought to have picked up the trick of washing by a process known as stimulus enhancement. This means that one individual has its attention drawn to an object by another's behaviour, and eventually stumbles across the same behaviour itself.

In the Taï forest, mother chimpanzees make use of such effects to help their infants learn how to crack nuts, using stones as hammers and tree roots as anvils. They leave hammers on anvils, or nuts near anvils. They also assist their infants by providing them with nuts, or good hammer stones. During sixty-nine hours of observations, Christophe Boesch and his colleagues recorded 975 such episodes. They saw teaching occur just twice. In one case, a chimpanzee known as Salomé repositioned a nut on the anvil for her son Sartre; in the second, a chimp called Ricci watched her daughter Nina struggle unsuccessfully with a hammer for eight minutes, then took the stone and very slowly rotated it until she was gripping it in the proper position, whereupon her daughter followed suit. Boesch believed these to be the first instances of teaching to be reported in primates other than humans.[77]

It seems reasonable to suppose that middlebrow Acheuleans had much greater teaching capacities than lowbrow chimps. If so, making handaxes might make an Acheulean female a better mother. She would be able to nurture her children's knapping skills, which would improve their sexual prospects, and thereby further her own genetic interests. These would have been considerably clearer than those of the males in the group, because of the ever-present uncertainty that hangs over fatherhood. Since a male could never be fully confident that he was the father of a particular juvenile, his interest in investing in the youngster's education would remain equivocal. Older males would probably devote some time to fostering the knapping skills of their juniors, as part of the weave of male coalitions, but the strongest currents of knowledge flowing between the generations would have been from mothers to sons.

Allowing that Acheulean hominids were in some sense human, we can attribute a recognizably human richness to the communication that surrounded their tools and their learning. The differences in complexity between chimpanzee and Acheulean learning would surely have been at least as great as those between nut cracking and handaxe-knapping. It is not easy to grasp the basics of fracture mechanics just by watching. The ability to teach may have been critical to the spread of artefacts made of flaked stone.

Acheulean females might also have benefited in exactly the same way as the males, since females were probably subject to sexual selection pressure too. Under a mating system where females are monopolized by a single male, the successful male does not need to be choosy once he has established his status.

While defending his position from other males may be costly, the cost of actual matings will be low. Under the complicated systems that are familiar among modern humans and are likely to have governed our hominid ancestors' reproductive fortunes, most males have opportunities to reproduce, but these are constrained and relatively expensive. Given that their opportunities are limited in practice, males should be inclined to exercise a degree of choice in making the most of these opportunities. Although modern human prejudices may make it hard to imagine that a male should be attracted to a female by her handaxe, there's no reason why a signal for the goose shouldn't be a signal for the gander.

The pressure to produce such signals would be less intense for females, though, while the pressure on them to secure food would be greater, since they would often have other mouths besides their own to feed. Geoffrey Miller summarized the sexual balance of interests thus: 'In females, the marginal costs of sexually-selected traits will be higher (because the demands of maternal investment push females closer to their physiological limits), and their benefits will be lower (because males are less choosy), so females often invest less time and energy in growing and displaying such traits than males do.'[78]

There are also instances in which females invest more in sexually selected traits than males. In some species of seahorse, the females are more brightly coloured, and put more effort into their courtship displays. The males, however, are the sex which invests more in reproduction; they take care of fertilized eggs by carrying them in pouches. Therefore they are exceptions which prove the rule.

A compromise proposal, taking account of the various forces in play, might run thus: females knapped flake tools and handaxes, with practical considerations to the fore in each case. Conversely, when they cast their eyes over the handiwork of the males, they were most impressed by the least practical artefacts – as long as the knappers proved able to take care of their own interests despite their self-imposed handicaps. In an assemblage of hand-axes, the less fussy, more sensible bifaces were more likely to have been made by females. Males were responsible for the extremely large, extremely small, and extremely symmetrical specimens. The maker of the Furze Platt Giant, who knapped a handaxe of dimensions suitable for somebody twelve feet tall, was signalling that he was twice the man he appeared to be.

Males might also have been responsible for the practice, noted by John Lord as he knapped in front of Steven Mithen's students, of chipping flakes off a surface which had itself been formed by striking off a flake. This improves neither symmetry nor service-ability, but as Lord remarked, it does counteract the impression that 'the stone's done half the work'. He added that this was a habit of both contemporary and Palaeolithic knappers. It would appear that in this respect, ancient and modern minds work alike.

Unless a few complete archaic hominid skeletons are unearthed with handaxes in their hands, the proposal of sex dif-ferences in handaxe technique will remain unsullied by direct evidence. It may, however, have a modest ability to play the sci-entific game of predict and test. Being more concerned than males to keep costs down, females would have used their tools more often before discarding them. The model therefore predicts that if handaxes from the same assemblage and of similar size were compared, the less symmetrical ones would show greater signs of wear. If any flake tools present had also been more heav-ily used, this would be consistent with the hypothesis as well. Evidence from flakes would also discriminate between the sexual selection model and functional hypotheses. A functional hypoth-esis would predict that as bifaces were more efficient, they would be more heavily used.

After working out what the sexual selection model would pre-dict about flakes and bifaces, I was encouraged to read a paper describing a particular spot at the Boxgrove site in detail.[79] Like the horse butchery site, its excavators interpret it as a place where a large animal's carcass was 'processed', though the bones themselves are missing. Eight handaxes were found, none of them showing any signs of wear; but there were fifteen flakes which had been damaged by use.

If the model were to be investigated by reference to the archaeological record, comparisons between the flakes and the bifaces within assemblages would be one of the most obvious lines of inquiry. The trouble is that the prediction is not properly reversible, because the costs of the two tool types are not equal. An assemblage in which flakes are more heavily used than bifaces is not a problem: it supports the model unequivocally. One in which the handaxes were more heavily used than the flakes is consistent with functional hypotheses. It does not support the sexual selection model, but unfortunately may not knock it down either. Since flake tools are cheap to make, they may have been disposable. A hominid could throw one away as soon as it began to lose its edge, and knap another one.

Without properly discriminating tests, any supporting evidence gained by field or experimental studies may be rather like the meat female chimps obtain by sexual exchange: a welcome bonus, but never enough to be the principal means of support. The model will have to rely upon its own strengths, to persuade rather than to prove. But as the model is made stronger, it should become easier to check against reality. We'll come back to this in a while.

Resuming the story so far, and adding some elaborations, this is how it runs:

Two and a half million years ago, hominids learned how to knap stone. Around this time, hominid brain sizes began to increase, but the stone tools remained essentially unchanged for about a million years. Members of the *Homo ergaster* species then began to make bifaces in crudely standardized forms, the Developed Oldowan industry. They did so because the new arte- facts offered some practical advantage; and they may have been under some environmental pressure to improve the efficiency of their tools. The climate was becoming more arid during this period, making food resources patchier.[80]

At first, the new tools need not have required any increase in cognitive capacity. It could have been little more than a matter of checking for V-shapes. As an edge is worked, the knapper needs to monitor its progress by looking at it side on, in which view it will form the apex of a deep V. Adding a check for a V shape in the flat plane would produce a tool with the basic triangular shape of the handaxe. Over a period of 100,000 or 200,000 years, this pat- tern was refined into the Acheulean industrial form.

This development has the hallmarks of sexual selection. A roughly triangular biface is more efficient than simple flakes or choppers, so individuals making bifaces are demonstrating adap- tive abilities. The biface is a highly visible indicator of fitness, and so becomes a criterion of mate choice. As resources become patch- ier, competition between males tends to increase; under the pressure of sexual selection, the bifaces' characteristics are rapidly exaggerated, and the Acheulean handaxe form crystallizes.

Selection revolves around the artefacts' symmetry, for which hominids already have a perceptual bias. Symmetry is often at a premium in sexually selected ornaments, from beetle horns to deer antlers, because symmetry is an indicator of fitness. Deviations from symmetry may indicate flaws in development, and evolutionary psychologists have claimed that modern humans are attuned to them: the more symmetrical a face, the more attrac- tive it is considered; the more symmetrical a man, the more sexual partners he is likely to have. Be that as it may, a perceptual bias towards the detection of symmetry is neither unusual in the animal kingdom nor cognitively demanding, and it makes selective sense.

Developed to the Acheulean stage, bifaces give diminishing

performance improvements in return for the extra costs of making them. This, however, renders them eminently suitable as indicators of fitness according to Zahavi's Handicap Principle. Fitness is indicated by the ability to sustain the additional costs of making a handaxe, without hindering the success of other activities. The handaxe also indicates fitness more directly, since it requires a combination of highly adaptive qualities: good physical condition, motor skills, eyesight, spatial perception, and above all, the ability to conceive and realize a plan.

In the phrase coined by Richard Dawkins, handaxes are part of an extended phenotype. The phenotype is the individual produced by the interaction of its genes and its environment; the extended phenotype is one in which the reach of the genes is extended using material that is not part of the individual's body. Although these are the human speciality *par excellence*, other species have produced dramatic examples. By the standards of the beaver, which builds log dams, or the caddis fly larva which builds itself a house of stones held together with cement, handaxes are not such a big deal.[81] It should not strain the imagination unduly to entertain the prospect that they played a direct role in the replication of hominid genes, as a mate choice signal, as distinct from a general role in promoting their makers' welfare.

In Zahavian terms, they correspond to the zebra's stripes or the eyes on a peacock's tail. An individual has to display an indicator which conforms to a standard pattern. Its observers have to be able to compare like with like. 'The markings that we see as *uniform* are the very ones that show most clearly the fine *differences* among individuals regarding the attributes most important to them,' the Zahavis observe.[82] Convention requires millions of men to wear suits of almost identical cut, in a narrow range of grey and dark blue shades. These garments can tell us little about character, but are excellent clues to wealth and social status. Thus Armand Marie Leroi was impelled to introduce his review of *The Handicap Principle* and W. G. Runciman's *The Social Animal* with a

paean to the suit Lord Runciman wore when he addressed a
Darwin Seminar (with the message that 'Social Darwinism is
wrong because neo-Darwinian sociology is right'). Runciman 'was
not dressed as were the other academics present,' noted Le Roi,
'what with the faultless lines of his suit, the softness of the fabric
(visible at 40 feet) . . . Quietly and unmistakably, that suit spoke
of money and influence.'[83] Specifically, it spoke of Runciman's
position in the City, as chairman of a shipping group.

An Acheulean hominid would have viewed another hominid's
handaxe with an equally discerning eye. As well as being the
Acheulean signature, the handaxe was the Acheulean suit. Like
the host of suits in the City's streets, handaxes tell us that their
owners needed to conform. So does the way in which assemblages
of them are dominated by a particular local style; ovates in the
case of Boxgrove. But they do not deny that their makers could
have been individualistic or creative in less visible ways, just as
City gentlemen, when away from their offices, may build model
boats, play the clarinet, or pursue the synthesis of sociology and
evolutionary theory. By establishing mate choice on a foundation
of reliable indicators, handaxes could have complemented
courtship procedures based on creativity and novelty, of the kind
which Geoffrey Miller has proposed as a driving force behind
human mental capacities.

Handaxes also fit at least two, and perhaps three, of the four
interpretations that Alan Grafen derived from Amotz Zahavi's
early writings. They are not unavoidable burdens, like the peacock's tail, which the bearer must endure in order to survive.
That, however, is the earliest and least subtle version of the
Principle. Handaxe-making affords an ideal opportunity to assess
the makers' quality, every bit as effective as watching them run a
race. Handaxes could therefore serve as revealing handicaps. It is
also possible that handaxe manufacture was beyond the capabilities of some Acheulean hominids, in which case the handicap
would to some degree have been condition-dependent. But the

version of the Handicap Principle to which they are most obviously suited, however, is strategic choice. If even a fraction of the Zahavis' inventory is accurate, the vast majority of handicaps operate in species with no insight into their operation. Strategic choice requires a higher degree of intervention by the individual than the other classes of handicap, and so might be expected to reach its most complex forms among species with highly developed cognitive capacities. Hominids, their intelligences driven by social dynamics, would be particularly well equipped to make fine judgements about how much of a cost to incur, and what the likely effect of their outlay on their companions would be. They would be able to suit each artefact to the circumstances in which they made it, and thus refine strategic choice into tactical choice.

The form of the handaxe lends itself superbly to sliding scales like these. When I raised the question of how fast a handaxe could be knapped, John Lord chose a flat piece which required relatively little working to turn into a finished biface. When Phil Harding opted to 'take something which is a bit more challenging', he put aside a similar stone and picked up a small boulder. Back in the Lower Palaeolithic, a male hominid of mediocre quality might consider which of his companions were on the scene before beginning to make a handaxe. If a better knapper was present, the average hominid could decide to assert that he was a player in the game by making a basic handaxe, but not to spend a great deal of time improving it if it would be outshone by the work of his more adept rival. Whether females were present, and which ones, would also be factors in calculating how much effort would be worth while. The average Acheulean male would be weighing up two questions: 'Who am I trying to impress?' and 'Can I be bothered?'. In this respect, Acheulean hominids were human as we know it all too well.

Acheulean industries may have exerted a degree of selective pressure upon technical intelligence. It's possible, however, that the really significant effect was upon other artefacts. The Schöningen spears, and the other wooden objects found with them, showed that archaic hominids were capable of sophisticated functional design, and suggested that they made much more than handaxes. If sexual selection honed the skills needed to design and execute handaxes, these abilities might have combined with other developing cognitive capacities to produce a range of specialized functional designs.

At the same time, the designs of the handaxes themselves would have been held constant by their function as fitness indicators, which required that like was compared with like. A remarkable possibility opens up: that the archaeological record has caused us to read the Lower Palaeolithic backwards, as a period of immense technological conservatism, when in fact the forces maintaining that conservatism also allowed a hundred new flowers of design to bloom.

Whatever was in the tool-kit, both sexes needed access to it. While the Acheuleans' artefacts were closer to those of modern humans than to those of modern chimpanzees, with their societies it was the other way around. Females needed to use tools in order to secure resources not just for themselves but for their offspring, since the sharing of food was a marginal activity at best. Operating under less sexual selection pressure than males, but more resource pressure, females tended to make more utilitarian artefacts. Experience of handaxe-knapping may have been useful in assessing male handiwork, however. Since females shared the genetic benefits of the selection pressure they exerted upon males, they would have had the cognitive wherewithal to make handaxes if they wanted.

For hominids in groups, making handaxes and making assessments of handaxe-makers were both demanding tasks. A hominid needed not just a technical perspective, but also a well-developed faculty for social orientation. He or she had to keep track of the other individuals in the group, and the interactions between them, to see how they coped with the handicap of company. At the same time, he or she had to maintain equilibrium in the seething waters of an intelligent primate group. Butchery events were among the most highly charged social situations, especially when the carcass was acquired unexpectedly, and handaxes had to be knapped on the spot. Such occasions would have stimulated individuals to produce the most impressive threat displays of which they were capable, to maximize their share of the meat. These displays would have been elaborate as well as vivid. Combined with biface knapping, which would also be driven to elaboration by the competitive pressure, they would constitute a pre-symbolic ritual.

12

If sexual selection is the key that unlocks the mystery of the handaxe, why has it been overlooked by the many scholars who have considered the problem? One answer is that the Handicap Principle, which integrates the message of the handaxe into the hominids' behaviour as a whole, is effectively a new one. Although Amotz Zahavi first promulgated it in 1975, it only began to win acceptance after Alan Grafen gave it mathematical backing in 1990. It has also fallen into a sectarian divide between those who believe sexual selection is about the identification of good genes, and those who think sexual signals are largely arbitrary, but are favoured because they make their possessors more attractive. The Handicap Principle is now the hottest game in the good genes camp, while the 'good taste' faction upholds the older tradition of Ronald Fisher. But Fisherian processes and mechanisms for indicating good genes are not mutually exclusive, as Malte Andersson points out in a comprehensive review of sexual selection theory, and the two forms of sexual selection may act in concert.[84]

Another answer is that the use of Darwinian models in archaeology is still at an early stage, and they have yet to be applied wholeheartedly to questions of mind. The terrain of ecology and body form is much firmer ground, so evolutionary theorists have concentrated on anthropology and fossils rather than archaeology and tools. In recent years, handaxe studies have tended to concentrate on the influence of raw materials upon the kind of artefacts produced. The authors of one paper refer explicitly to sexual selection – but only as a rhetorical device, turning on the opposition between artifice and nature. Assuming that natural selection works on bodies, while minds work on artefacts, Bruce Bradley and C. Garth Sampson dismiss the answer while pointing straight at it. '[W]e know (some of us occasionally forget) that variability in Acheulian artefacts was not generated by natural and sexual selection,' they write, 'but by a set of very different factors.'[85] Yet both the parenthesis and the chance remark by the archaeologist at Boxgrove, which set me off on the sexual selection path, shows that the ball was already in the archaeologists' field of play, waiting for someone to run with it.

And in fact one scholar did, during exactly the same period that I was thinking out my model. Independently, Steven Mithen and I both realized that sexual selection could be the answer, operating upon criteria of symmetry.[86]

Steve had worked on the handaxe problem before, trying to answer the Clactonian question: why some groups of hominids in southern England had left behind Acheulean assemblages, while others living at roughly the same time had left so-called Clactonian assemblages, containing few handaxes or none. He had identified the landscape as the factor that made the difference. Most assemblages lacking handaxes were produced in woodland areas, while Acheulean assemblages were made by groups living in open country. In the woods, animals that can climb have the trees to help protect them against predators; in the open, they have to rely more on each other, and so tend to congregate in larger groups. The larger the group, the more intense and complex are the interactions between individuals within it. The conditions are stressful, and in that respect costly, but they favour 'social learning'. Individuals pick up the behaviour of others more readily, being more frequently exposed to it. At the same time they are constrained in their activities by the cognitive demands of life in a large group, reducing their opportunities to experiment, or to learn by trial and error. Mithen therefore reasoned that Lower Palaeolithic hominids would be more likely to learn how to make handaxes in a large band, and less likely to develop designs of their own. In small groups, the conservative weight of handaxe culture would be lifted, and tools could be made outside the dictates of the pattern.[87] The sexual selection model added another leg to this argument. In smaller groups, competition for mates would be less intense. It would simply not be worth a hominid's while to go to the trouble of making a handaxe.

While I had been concentrating on the costs, Steven Mithen had also been pondering a couple of very pertinent issues which I had not addressed. One was that of cheating. What was to stop inferior but sly hominids from picking up handaxes and passing them off as their own work? Mithen had realized that here lay the answer to the striking archaeological phenomenon of handaxes in large numbers and almost pristine condition. The absence of visible wear and tear does not prove a handaxe has not been used. It may have been employed solely on flesh or soft vegetation, or its edge may have been touched up to resharpen it. Nevertheless, it is striking that none of the handaxes found at the Boxgrove site have edges damaged through use; and Boxgrove axes are not exceptional in this respect. Observers could only be sure that the wielder of a handaxe was also its maker if they saw the axe being made. After use on the spot, or for a short period afterwards, it would have to be discarded. A handaxe could only be trusted if it had barely been used.

This insight made a perfect match with my ideas about handicap. Discarding handaxes immediately would increase their cost, making them items of even more conspicuous consumption than before. Once a handaxe had been made, in any case, its job was largely done.

The practice of immediately discarding handaxes would not entirely be proof against cheating, but that would not matter. If adhered to strictly, it would require successful cheats to have much the same qualities as expert knappers. Cheats would have to be able to engineer situations in which they knapped undistinguished bifaces, then substituted fine handaxes without onlookers realizing. The larger the group, the more pairs of eyes there would be to evade, while the costs of countersurveillance to each individual onlooker would remain very low. A successful cheat would therefore need to have basic knapping skills, including the ability to secure a share of resources while making a handaxe, exceptional Machiavellian skills, and, just like an accomplished handaxe-maker, a highly developed ability to conceive and execute a plan. In sum, a cheat would need not only technical and social intelligence, but also determination and, if detection were to result in punishment, a certain amount of courage. He would therefore be a good catch for a female in search of a mate: his devious ways would matter little to her, since she would want little from him besides his genes. A stringent system of measures against cheating would make successful fraud an honest signal of quality!

Sanctions against cheating raise a question fundamental to any society with human qualities: Were there rules? Perhaps that should be rephrased. Were there Rules with a capital 'R'? In some sense, there must have been rules, but that sense may not be interesting except in engineering terms. There might simply have been a final line in the knapping program which said 'now throw it away', and a counterpart in the mate choice program that said 'go for the ones that throw them away'. These instructions could have been selected automatically, like courtship rules in grouse or bowerbirds. But knapping handaxes is not an automatic procedure. It is a complex form of behaviour that draws upon cognitive capacities available only to hominids.

There is video evidence that bonobos may be able to make Rules. Members of a wild group living at Wamba, in Zaire (Congo), appeared to operate a rule that, on arrival at a site where they were given sugar-cane by their human observers, they had to wait for a signal from their alpha male before they could start to eat. One video sequence from the study shows a juvenile jumping the gun, his eye on the alpha. An infant standing beside the alpha male starts to make excited noises; the alpha stands upright and raises his right arm. In a submissive manner, the juvenile approaches the alpha, who steps on his back. The juvenile then withdraws to the margins of the group, where he awaits permission to take part in the communal feeding.

Barbara King and Stuart Shanker describe this incident in a discussion of what constitutes a rule.[88] A statistically predominant pattern of behaviour is not sufficient to demonstrate the operation of a rule, they point out. 'To claim rule-following,' they continue, 'we need evidence of instruction directed to young or new members of the group about the nature of the rule and when and where it applies, as well as evidence of sanctions against those who break the rule, and perhaps even awareness on the part of the rule-breaker that a rule has been broken.'

The video is admissible as evidence for sanctions, and possibly of awareness. But it does not demonstrate instruction, nor prove that a rule is in operation. The juvenile may appreciate that taking food before the alpha's cue is likely to provoke the alpha to assert his dominance, just as a child may learn that certain acts provoke unpleasant assertions of parental power. Whereas the child will eventually induce the existence of rules from its experiences, what the young bonobo is able to learn from his punishment remains unclear. Applying the rule of thumb that Acheulean hominids fell somewhere between chimpanzees and living humans in their capacities, if it is possible that bonobos can make rules, it is probable that Acheuleans could. As well as the means, they had the motive; a sexual one, as it so often is.

If the pressure to police knapping displays did result in the establishment of a rule about the discarding of finished hand-axes – or if rules arose to govern any of their other activities – then the Acheuleans had one of the defining elements of human society. They need not have had any deep awareness of what the rule was for, however. There was no need for them to be consciously aware of the connection between the quality of a handaxe and the fitness of its maker. Modern humans have managed quite effectively without understanding why certain traits appeal to them, if the evolutionary psychologists are right. This isn't an especially remarkable view. They are only applying the fundamental axiom of psychology, which is that we don't really understand the reasons for our actions. If we did, there would be little of interest for psychologists to do.

There is certainly no need for modern humans to know that symmetrical features may indicate that their owners have enjoyed a steady passage through their early developmental stages, and are therefore likely to have a sound set of genes. All they have to do is find symmetrical features attractive. As intelligent primates, the Acheuleans may have noticed that individuals who made impressive handaxes tended to be physically impressive themselves. Those who made this connection would have been at an advantage, since they would have smelt a rat if they saw a fine handaxe in the hands of a poor specimen. But the Acheuleans would not have needed to understand that handaxes were reliable indicators of fitness. All they needed to understand consciously was how to make a handaxe.

The mind of an Acheulean knapping a handaxe would have worked in ways that had much in common with that of a modern knapper, given the demands of the task for planning and spatial manipulation. Outside this slice of behaviour, the Acheulean mind would have moved in radically different ways. As far as the handaxe handicap system of sexual selection was concerned, only a technical consciousness was required, and the most economical assumption is that the Acheuleans were no more conscious than they needed to be. As they knapped their rocks, the Acheuleans were human as we know it. As they chose their mates, they were not.

13

Everything comes to an end, even the immensely stable Acheulean industry, but why did it break down when it did? If it arose because of sexual selection, then it probably ended because of a change in the relations between the sexes. This makes sense in terms of handicap theory, which applies most clearly to heritable traits. In other words, if all that males contribute to their offspring are genes, then females will choose mates according to criteria indicating genetic quality. The more that males are involved in the rearing of young, the more that females will make choices based on indications that males will be good providers. And the more that it takes to rear young, the more interest females will have in securing the assistance of males. The period during which the Acheulean industry began to break down, 300,000 or 250,000 years ago, was one in which brain sizes were increasing rapidly, and therefore the energy requirements of young hominids would also have been rising significantly. Robert Foley identifies this phase as a turning point in human evolution, at which hominids stepped on to the path that took them to an essentially modern human life history, with the phases of life from gestation to adulthood arranged in the proportions we know today.[89]

Foley points out that one way to lighten the energetic load is to spread it over a longer period, and that as hominids have evolved they have taken longer to grow up. This reduces the number of offspring born, and so increases their value. 'It is here, perhaps, that the strategy of the male may change,' Foley observes. 'To ensure the survival of these costly offspring a greater level of paternal effort may be advantageous . . .' As he and Phyllis Lee showed in their computer modelling exercises, males probably did not make much of a contribution to the costs of raising off-spring in the early stages of hominine evolution.[90] The change in male strategy is more likely to have taken place during a late stage of brain expansion, and the one which coincides with the end of the Acheulean era is an obvious candidate.

Under such conditions, females might have begun to choose males who were prepared to invest, rather than to spend conspic-uously. They would have been impressed by males who could acquire resources most efficiently, rather than those who could sustain the greatest handicap. Males would have had to redesign their tools for maximum productivity. As females ceased to select mates by comparing their artefacts according to the standardized handaxe patterns, the grip of the Acheulean would weaken. Although bifaces would persist in forms that were practical for particular tasks, the pressure to provide resources would result in the development of an unprecedentedly diverse tool-kit, with artefacts designed for different purposes.

Robert Foley believes that the key shift in technology at this stage, towards the use of prepared stone cores, was directed towards the use of projectiles and the improvement of hunting performance. It could thereby have increased the male potential for provisioning. The end of the Acheulean could be conceived as the first great industrial rationalization, eliminating the social dimension of manufacture in favour of productive efficiency.

14

At this point the proper question to ask is, do I really believe all this? The answer is yes and no – and much of the model's value lies in the weighing of this balance.

When I look at a handaxe, I do not feel certain that its form was the result of sexual selection, maintained over a million years. I find it hard to imagine how such a process might look and feel. And I'm not the only one. When Steven Mithen first outlined his ideas about handaxe symmetry to his Cognitive Archaeology students, they seemed to suspect that he was suffering from a slight touch of madness. Other people, scholars and lay, are likely to dismiss this model as absurd.

There are good and bad ways of objecting to the model. Objections based on the quality of the evidence or the logic of the argument may be good ones. Objections based on the feeling that the idea simply seems peculiar are poor ones. They derive from unfamiliarity with evolutionary theory, and an inadequate imagination. Many highly educated people suffer from the former. When it comes to prehistory, everybody is handicapped by the latter. It is hard to imagine hominids choosing mates by watching the opposite sex make handaxes. It is also hard to imagine a world full of trilobites, but we know that such a world once existed.

We know that trilobites evolved, too: for those of us who recognize, with Pope John Paul II, that evolution is 'more than just a hypothesis', evolution is as true as anything which we cannot actually hold in our hands. However, many people have found their imaginations insufficient to support this insight, and have used this inadequacy as the basis for arguing against evolutionary theory. When scholars say 'I just don't believe it,' they may think they are saying 'I speak with authority and dismiss it as absurd.' I suggest that, where modern evolutionary ideas are concerned, what some of them may really be saying is 'I just can't see how it could be that way.' Richard Dawkins has labelled this the 'Argument from Personal Incredulity'. One thing we may be confident about, however, is that Acheulean hominids must have done many things that would seem stranger to us than using handaxes as a factor in their choice of mates.

The strangest thing that we know for certain they did was to make handaxes for a million years, and to scatter them across three continents. If sexual selection played no part in maintaining this tradition, then we have to suppose that imitation and teaching were forces sufficiently strong to spread handaxes from South Africa to southern England. There are two ways in which they could have acted. Either they upheld a single unbroken tradition, or handaxes were invented many times and sustained by local processes. The latter would imply that the handaxe was such an overwhelmingly advantageous design that where it didn't exist, it had to be invented. Yet that does not seem to have happened across eastern Asia, which appears to have been settled by hominids who left Africa before the handaxe emerged.

The sexual selection model does imply a single tradition, but proposes a stronger mechanism for maintaining it. Instead of depending solely upon imitation and teaching, tradition is upheld by these two processes acting in concert with sexual selection. Faced with the extraordinary facts of the handaxe's range, this seems the least improbable explanation.

There are, however, plenty of good reasons to criticize the model. The main one is that it is a solution to a problem that doesn't exist. Steven Mithen and I were both impelled by the sense that despite all the conventional explanations, there is still a question mark over the handaxe which refuses to go away. The alternative position is that there is nothing particularly odd about handaxes. The form was serviceable, it offered some improvements over unstandardized tools, and so it continued until hominid faculties improved. Academic papers are not usually explicit about the last of these assumptions, but the underlying attitude is recorded by the Leakeys' biographer, Virginia Morrell. In her book *Ancestral Passions*, she remarks that Mary Leakey 'would refer to *Homo erectus* as that "dim-witted fellow" – annoyed by his apparent one-million-year obsession with fashioning the same style of handaxe over and over again'.[91]

In other words, handaxes are boring, and they require only a boring explanation. Their makers were, if not hewers of wood, mere hackers of meat. This is an entirely respectable view. It may even seem obvious. But there is often more to the work of primates than at first appears. It might seem obvious that chimpanzees hunt in order to obtain food. Under close observation, however, the decisions chimpanzees made at Gombe turned out to be subtle ones, whose patterns suggested that chimpanzees hunt in pursuit of sex as well as nutrition. If hunting can occupy both the sexual and the nutritional dimensions of chimpanzee life, it should not be so strange to imagine that artefacts could occupy the sexual as well as the material dimensions of middle-brow hominid life. They do that among the living hominines, after all.

In deciding whether there is an Acheulean problem or not, we're faced with several possibilities. One is that a better tool design and a dim wit account for the Acheulean entirely. A second is that sexual selection is the secret of the handaxe, as I have argued here.

The thing that allows the sexual selection model to hold its head up in scientific circles is that it points to a limited programme of tests. These may allow the balance to be weighed between it and the functional accounts. As presently constituted, it stands or falls on costs. We now need figures, rather than impressions. Anna Barker, one of Steven Mithen's students, began the process by studying replica handaxes made by John Lord. She concluded that handaxes are symmetrical to a degree which cannot be explained as a by-product of the knapping routine, or by functional considerations.[92]

Unfortunately, we will not be able to feed a set of data into a set of equations and thereby settle the matter. For one thing, the extra costs of handaxes may have varied during the course of the Acheulean era. The first hominids to make handaxes were very different from the last ones. They had considerably smaller brains, and they may have found it far harder to impose a form on a piece of stone than their distant descendants. Whether the benefits of handaxes were functional or sexual, they presumably enhanced the reproductive success of their makers one way or another. A million years of the consequent selective pressure would almost certainly have honed the skills needed to make them. Modern experimental knapping would therefore be a much better model for late Acheulean time economies than for the efforts needed to make the earliest bifaces.

A more serious problem arises from the fact that the true costs depend on the context. If Acheuleans could fashion stones at their leisure, it would not matter that a biface took so many per cent longer to complete than a flake tool. The sexual selection model proposes that they were made under conditions where time was of the essence, the Acheulean rules requiring that handaxes be made when needed, not prepared in advance, and then discarded afterwards. In this account, I've built the argument on the slender platform of a single find at Boxgrove. It would be strengthened by other cases in which knapping *débitage* is associated with butchered bones, but guilt by association is not proof. The bones and the tools could have been deposited centuries apart.

A more systematic approach to the problem would involve surveying the archaeological record to get some figures on the extent of phenomena that we have used in building the argument. It would be useful to know more about how often extremely large, small or highly worked handaxes are encountered. If they are found mainly where the stone lends itself to knapping, we may need to identify what would constitute an extreme handaxe in areas where the rock is less co-operative.

Olduvai is an interesting site in this respect, since several different types of stone were used. In his replication experiments, Peter Jones learned the different properties of each. The most amenable was a green lava called phonolite, which 'easily accepts many sorts of subtle flaking techniques of the kinds that are readily applied to English flint'.[93] Jones noted that 'these qualities can be seen to have been used to their limit', and indeed some of the results are exquisite. In the terms of the sexual selection model, the hominids added value where they could.

15

Steven Mithen was not the first archaeologist to use the hand-axe to make a model of ancient hominid minds. In the late 1970s, Thomas Wynn began publishing papers which placed stone tools in the framework built by Jean Piaget, whose influence on ideas about how children's minds develop is comparable to that of Freud upon psychopathology. Wynn pointed out that Piaget himself would have rather dealt with *Homo erectus* than children, quoting the master to the effect that 'the most fruitful, the most obvious field of study' would be 'the history of human thinking in prehistoric man'. Specimens were regrettably unavailable, but there 'are children all around us'.[94]

Piaget's theories are based on a model of the mind as a single engine, of general purpose, rather than as a complex of specialized subsystems. All the mind's abilities therefore grow up together, as the child reaches successive stages of development; all are based on a universal code of logic and mathematics. From birth to the age of two, the child occupies the sensorimotor stage, distinguished by characteristic infantile actions such as sucking, grasping, kicking and throwing. These actions can be combined into sequences, but cannot be represented in thought. At two years, the pre-operational stage begins, permitting the internal representation of action sequences. The child acquires language, uses mental imagery, and can focus on one aspect of what it perceives, such as height or colour. Around the age of seven, the child attains the stage of concrete operations. Numbers and logic become accessible, and the child becomes able to consider more than one aspect of what it perceives at the same time. This allows it to grasp that if water is poured from a tall thin glass into a short wide one, the amount of water remains the same. Finally there is the stage of formal operations, not reached until the child is twelve years old, in which thinking systematically and abstractly is possible.

Wynn argued that Oldowan tool-makers got by with simple concepts, such as proximity – in practical terms, aiming a blow next to the scar left by the previous one. They only needed to consider one effect of their actions at a time, and therefore required only pre-operational intelligence, a level attained by living apes.[95] The refined handaxes of the later Acheulean were a taller order. Before striking a blow, the knapper had to consider its effect on symmetry in three planes. Handaxes demanded the operational stage of intelligence, and the later Acheuleans were therefore not significantly less intelligent than modern humans. If intelligence had remained on a plateau for the past 300,000 years, the sophistication of artefacts created in the last 30,000 years might be due to the development of culture, not of innate intelligence.

John Gowlett thinks that Thomas Wynn may have underestimated what it took to make Oldowan tools, by concentrating on the positioning of blows, at the expense of the skills needed to make them effective.[96] Kanzi the bonobo appeared to have mastered the requirement of proximity, directing his blows to the same general area of the rock, but he had not grasped fracture mechanics. Like a novice human knapper, his efforts typically produced flakes that were very small or had blunt edges. An intuitive understanding of stone fracture may be what separates hominines from apes, at least as far as tool-making is concerned.

Even if this distinction is granted, Wynn's scheme of intelligence does not sit comfortably with current ideas about cognitive evolution. Since his early formulations were published, a growing body of experimental evidence has challenged Piaget's conclusions about how young minds develop. According to Piagetian theory, for example, infants take a year or so to work out that an object continues to exist if it disappears from view. Experiments conducted in the 1980s indicated that babies realize this at the age of three or four months.[97]

Research of this kind has promoted the concept of mental 'modules', whose rise can be conveniently dated from the appearance in 1983 of Jerry Fodor's book *The Modularity of Mind*.[98] Fodor argued that a mind based solely upon a universal engine – 'domain-general', in the jargon – would not work. It would be far too slow for the real world, searching through an indefinitely large space of thought for solutions to each new problem it encountered. While it was doing so, its owner would probably come to a rapid and sticky end. Organisms need fast reactions, but human individuals plainly benefit from the ability to reflect at leisure too. Fodor proposed that human evolution had managed to get the best of both worlds. He conceived of the mind as being split into perceptual and cognitive systems. Perceptual processing is conducted by discrete modules, each handling its own kind of sensory input in its own kind of way. A module cannot receive an input from another module. Instead, communication is exclusively radial, each module feeding its output to the cognitive core of the mind. Perception is fast, automatic and unintelligent; cognition is slow, often conscious, intelligent and reflective.

A modular scheme for the mind makes narrative sense. Animal behaviour suggests that animal brains comprise systems that are fast, specific and relatively unintelligent. As these organs are subject to natural selection, modern Darwinism presumes that their subsystems are adaptive. If a subsystem is specialized to cope with needs encountered by a wide variety of species – to assess whether an animal poses a threat, for example, or to pick out edible plants from a tangle of vegetation – then natural selection is likely to maintain it. Hominids would not lose faculties possessed by the Last Common Ancestor, unless new environmental conditions rendered them redundant, or other demands made greater claims on processing capacities. Although some faculties would surely fall victim to changes of circumstance, it seems reasonable to suppose that others would be conserved. Dan Sperber, an advocate of 'massive modularity', argues that modules might become more sophisticated, but they would not become more general. 'Loosening the domain of a module will bring about, not greater flexibility, but greater slack in the organism's response to the problem,' he observes. Evolution proceeds by single steps; the immediate effect of any single step towards generality would be counter-adaptive, and so it wouldn't happen.[99]

Fodor's original model, bold enough in its time, was mildly reformist compared with the florid modularity developed by the evolutionary psychologists. In making the case for dedicated modules around the periphery of the mind, he seeded the conceptual ground, but he retained an old-fashioned universal engine as the mind's core. According to his 'First Law of the Nonexistence of Cognitive Science', this inner mind is unknowable. When other theorists took his ball and ran with it, he dismissed their ideas as 'modularity theory gone mad'. Today, he remains scornful of 'psychological Darwinism'. He insists on calling its practitioners sociobiologists rather than evolutionary psychologists, and he finds their account of human nature 'grotesquely improbable'.[100]

In leaping forward but then declining to move any further on, Fodor resembles Noam Chomsky, who transformed the study of linguistics in the 1950s and 1960s with his theory of 'universal grammar'. Chomsky drew up the template for evolutionary psychology by pointing out the signs that indicate the existence of a dedicated mental organ. Children do not need to be taught to speak in the way that they must be taught to swim or do arithmetic. The languages that they speak share the universal structures that Chomsky described. It looks as though the word processor is included in the package.

Fodor and Chomsky share a readiness to accept that parts of the mind are not a blank slate, and a marked resistance to proposals about the influence of natural selection on the mind. They are nativists, but they are not adaptationists. Chomsky cannot see how evolution could have created the language organ, and regards it as a by-product of the human brain's massive expansion. Like Stephen Jay Gould, who shares his views on the evolution of language, he sympathizes with those who seek explanations for biological form in structure rather than selection.[101] Fodor cannot abide the 'stuff about selfish genes', of which he doesn't believe a word.

In the adaptationist camp, John Tooby and Leda Cosmides conceive the mind as one great mesh of modules, dedicated to tasks such as recognizing faces, using tools, being frightened, allocating effort, looking after children, perceiving emotions, imagining what is going on in the minds of others, conducting friendships, understanding the behaviour of rigid objects, and many more. Yes, they say, modules are better than general processors at handling tasks such as vision or grammar. So why stop there? Why should it not be modules all the way? Riding the wave of module logic, they are able to proceed by wish-lists, making cases for different faculties which would have been particularly advantageous for our ancestors.

For onlookers who do not feel inclined to buy into any one of these positions wholesale, the developmental psychologist Annette Karmiloff-Smith offers a way to reconcile modularity and experience. In her book *Beyond Modularity*, she sketches a synthesis between a partitioned and a general mind; between Jerry Fodor and her former tutor Jean Piaget.[102] Drawing upon experimental advances in the study of child development, she proposes that the mind contains a relatively small number of innate dispositions, which direct its activities towards particular domains. Modules are seeded by the initial dispositions, but they are built in the course of development, through the interactions of individuals with their environments. Since so many features of children's environments are universal, such as rigid objects or human faces, a universal human psychology could still emerge.

16

The sexual selection model places the handaxe in the realm of social relations, as well as that of practical tools. According to Steven Mithen, categories like these are not simply forms we impose on the ancient mind in order to make some sense of it. In his book *The Prehistory of the Mind*, he argued that the way the early hominine brain distinguished itself from its ancestors was by the development of modules dedicated to different aspects of its world. The mind was divided into compartments, with relatively few connections between them.

Mithen drew his reasoning from the practices of software developers, and his principal metaphor from the work of church architects in the first two-thirds of the Christian era. When developing a complex new program, he observed, software designers first have to produce a basic system that works. This corresponds to the basic early hominid brain, roughly comparable to that of the chimpanzee. It has general capacity, but little in the way of specialized subsystems.

In order to add complexity, the designers must work on the

subsystems separately, making sure they work properly before integrating them into the package, one by one. The mind of *Homo ergaster* or *erectus*, much larger than that of the earliest hominids, corresponds to this stage of development. A number of new systems have been assembled, but they have not been fully integrated into the whole, and communication between them is limited. If a divine Developer had set out to make the modern human mind, this is how the production process would have unfolded. The modules of the ancient mind were not created as part of a Sapiens Project, though. Each emerged as a response to selective pressures in different cognitive domains. Once they were in place, weaving them together became an evolutionary option, and it proved adaptive. Mithen's colleague Mark Lake has suggested that alternation between generality and complexity must be common in evolution, because what is true of software and minds is true of complex systems in general. In Daniel Dennett's phrase, it's a 'good trick' that will be repeatedly rediscovered. In the terms laid down by an earlier philosopher, Hegel, it's a dialectic that results in a higher synthesis.

For his central metaphor, Mithen recalled his student experience in 'dirt archaeology', excavating the South Church at the Abbey of San Vincenzo in the Italian town of Molise. Over a thousand years, he and his colleagues concluded, the South Church had undergone five architectural phases, passing from one to another in outbursts of wall demolition, floor-laying, doorblocking, and the addition of storeys. Archaeological inquiry entails the identification of these changes in the structure of the building. By studying the prehistoric archaeological record – neglected by evolutionary psychologists, as Mithen has pointed out – cognitive archaeology seeks to describe successive phases in the building of the mind. Mithen identifies three such phases.

The first is like a chapel built in the earliest Christian times, simple and modest in proportions. It is a unitary mind, with minimal subdivision. The Last Common Ancestor had a mind like this, as did the australopithecines. Modern chimpanzees still do. They rely mainly upon general intelligence, with perhaps a module specialized to cope with the demands of their social world, and micromodular maps to help them locate food. Their social intelligence module operates to a substantial extent in isolation from their general intelligence. Perhaps this, suggests Mithen, is why the chimpanzees of Taï so rarely teach their young to crack nuts using stones. A chimp can imagine what is going through the mind of another if the thoughts in question concern the social world, but not if they are about matters such as tool use.

As the mind's evolution continues, modules are added, like the chapels that appear around the nave in the Romanesque period. These chapels are largely isolated from each other; sound and light are confined within each. This is the High Fodorian phase of the mind's development, with modules dedicated to specific domains, operating in isolation. It begins with the construction of a rudimentary technical intelligence module, giving Oldowan hominids their insight into how stone fractures. By the later Acheulean period, hominids have developed their social, technical and natural history intelligences extensively, but they cannot bring each to bear on any other.

This needs to be qualified slightly in order to reconcile it with the handaxe handicap model, according to which handaxes are supposed to show a hominid's mettle in more than one domain. A knapper has to think about the raw material, about how to impose form upon it, and about how much effort to invest in making it. The latter consideration depends on the individual's assessment of what is going on in the group at that particular moment, and how those dynamics affect the individual's own interests. It implies a far more powerful intelligence than that needed to hunt deer or horses. At butchery sites, the knapper's decisions may need a modest input from the natural history module, factoring in knowledge about the amount of meat that a carcass is likely to yield. Placing intense demands upon both social and technical intelligence, while also involving natural history intelligence, handaxes would seem to demand precisely the kind of cognitive integration whose absence, according to Mithen, characterized the Acheulean mind.

The impression of interwoven domains is dispelled, however, by considering what each domain needs to know from the other. All that technical intelligence needs to know from social intelligence is how much effort to put into a handaxe. All that social intelligence needs to know from technical intelligence is how much effort went into the handaxe that is being assessed as a index of its maker's quality. In a model of the Acheulean mind, these outputs could be represented on a scale of one to ten. Once the technical module has received the instruction 'seven' from the social module – perhaps via the central engine devoted to general intelligence – it can compute for itself exactly how to allocate this level of investment. The knapping procedure is largely a closed one. Decisions about shape, size and quality of finish are made within the constraints of the bifacial form, the rock being used, and the knapper's skills; but not with reference to the details of the social context in which the handaxe is being made.

In practice, there would be a continuing series of instructions from social intelligence, but they would not need to express anything of the complexity from which they were generated. At a butchery session, for example, a knapper's social intelligence module might be capable of detecting that a squabble was brewing between two of the hominids, and inferring that this would allow two others, whose friendship the module had registered as stable, to collaborate in seizing a disproportionate share of the carcass. Sophisticated as this political and psychological analysis might be, the message it generates for the technical intelligence module is simply 'hurry up!'. Similarly, the input from the natural history module, advising on the amount of meat likely to be available from a carcass, could be reduced to something like 'small', 'medium' or 'large'. General intelligence would probably be sufficient to assess the rough yield from a rhinoceros or an elk, obviating the need for the natural history module to be involved.

In the final phase of Mithen's cathedral-building, the mind enters its Gothic phase. Church architecture is reconfigured around the circulation of sound and light. Though the chapels remain, they are now in mutual communion. One of the first walls to come down is that between the social and natural history chapels. Now people can make the symbolic associations between animals and humans implied by the presence of animal bones in the graves of early modern humans at Qafzeh and Skhul in Israel, or startling imaginative expressions such as the statuette found at Hohlenstein-Stadel in Germany, carved 30,000 years ago, which is half human and half lion. The technical intelligence chapel has not yet been opened up, however, which is why the early modern humans at Qafzeh and Skhul used tools similar to those of Neanderthals, and why they did not place artefacts in their graves along with jawbones and antlers.

When the wall surrounding technical intelligence comes down, the mind becomes fully modern. Annette Karmiloff-Smith envisages this condition as one in which knowledge belonging primarily to one part of the mind is redescribed for other parts, creating a series of representations. Two other developmental psychologists, Susan Carey and Elizabeth Spelke, think of it as 'mapping across domains'. Steven Mithen calls it 'cognitive fluidity'.

However much we work out about the early hominine mind, we will never really be able to imagine what it must have been like to be an Acheulean hominid, thinking in compartments. They themselves were probably unaware of the partitions within their minds, and likewise we are largely unaware of our own mental architecture. We think fluidly and in language; we cannot think through language to the underlying mental processes. It is so difficult to imagine thinking without language that many people believe languageless thought is not even possible.

What made language possible, though? The next section, 'Trust', examines one of the conditions that had to be met in order for humans to create language and culture, thereby developing a mind that is human as we know it.

THREE

Trust

1

Lower

At this depth, there are little more than scratches. Most of them are in the upper levels of the Lower Palaeolithic, the oldest division of the Old Stone Age. Faint lines are scored on pieces of rock or bone; roughly parallel, loosely fan-shaped, occasionally right-angled.

In 1980, at a site on the Golan Heights called Berekhat Ram, Israeli archaeologists discovered a pebble of volcanic rock, 35mm in length, that resembled a female human form. This resemblance was enhanced by two grooves,[1] one emphasizing the 'arms', the other the 'neck'. The archaeologist Alexander Marshack examined the grooves using an electron microscope, and concluded that they were made by hominid hand. Dating tests gave a range between 233,000 and 800,000 years. The object is likely to be more than a quarter of a million years old and has a claim to be the oldest known statue.

In Auditorium Cave, near the Indian city of Bhopal, there are nine round marks shaped like cups (these are known as cupules). Excavations in 1990 revealed a tenth cupule, and a meandering line pecked into rock which is covered by strata bearing Acheulean handaxes. Most Acheulean finds in India are over 350,000 years old. If these markings are deemed to be art, then they are the oldest art yet discovered on a cave or rock surface.

Middle

According to André Leroi-Gourhan, the most influential theorist of prehistoric art in the past fifty years, the caves he excavated from the late 1940s until the 1960s at the central French site of Arcy-sur-Cure housed 'the earliest of museums'. This was made up of 'a whole collection of "curios" – nodules of iron pyrites, fossil shells and fossil madrepores, which the men collected far from the caves of the River Cure and brought back to their dwelling'.[2] Nodules of iron pyrites contained shiny metallic flecks, the shells were spiral in form, and the madrepores, a kind of coral, were pitted spheres resembling golfballs. They are curios, to our eyes, and they were curated in the archaeological sense of having been collected and kept. It is a short step to the supposition that the hominids who brought them to the caves did so because they found them attractive or interesting. Judging by the tools found in the same levels, the first curators may have been Neanderthals.

At what is now Tata, in Hungary, a Neanderthal seems to have gone a step further. Once again, a fossil was involved; this time a nummulite, which takes the form of a disc. A natural fracture traces a diameter across its surface; a second line has been deliberately carved at right angles to it. The crossed circle is a figure to which people are perennially drawn, from Celtic Christians to the designers of icons that spin on a screen to indicate that a computer is busy. That covers fewer than 2,000 years, though, whereas the line was scored on the nummulite around 100,000 years ago.

Upper

A few days before Christmas 1994, Jean-Marie Chauvet and his fellow potholers, Eliette Brunel-Deschamps and Christian

Hillaire, discovered a cave near Avignon, in the southern French department of Ardèche. The Grotte Chauvet, as it was named, contained 300 animal images, including rhinoceroses, lions, mammoths, hyenas, bears, reindeer, ibex, two yellow horses, a red panther and an engraved owl. There was also a centaur-like composite of human and bison, some stencils of human hands, and an abundance of red dots, many arranged in geometric patterns or animal forms.

Grotte Chauvet is astonishing not just for its richness but for its antiquity. Over 30,000 years old, more than twice the age of Lascaux and Altamira, it is the oldest known example of cave painting. Yet the repertoire of techniques appears almost fully formed, except that the painters did not use more than one colour on a single figure, which is such a breathtaking feature of the paintings at Lascaux and Altamira. Images are superimposed and offset; the contours of the cave walls are used to add a third dimension to figures; surfaces are scraped away before the lines are drawn; there is shading and depth.

Although its concentration of variety is exceptional, Chauvet is characteristic of its period, in which art and technology suddenly flourish as never before. The desert blooms.

2

The three Palaeolithic levels, Lower, Middle and Upper, were defined on the basis of tool industries found in Europe. Only the Lower category is applied to African relics of roughly equivalent antiquity, the more recent of which are assigned to the plain English Middle and Later Stone Ages. Like the classes into which the hominids of these eras have been sorted, they are fuzzy and provoke disputes. Scientists are increasingly aware of their limitations. The Arcy-sur-Cure finds, for example, are associated with an industry called the Châtelperronian, a grey area which can be seen as the last of the Middle Palaeolithic or the beginning of the Upper Palaeolithic. The Châtelperronian is widely regarded as a stage in which some of the last Neanderthals raised their game as a result of contact with modern humans, enhancing their Middle Palaeolithic toolkit with improvements along Upper Palaeolithic lines.

Nevertheless, the habit of dividing prehistory into three parts will probably endure for a long time, being familiar and convenient. The Lower Palaeolithic is deep prehistory, more than two million years' worth of Oldowan and Acheulean industries. The Middle Palaeolithic is the stage at which tool-kits become more diverse, and is associated with Neanderthals. Stone blades are the hallmark of the Upper Palaeolithic, but the real significance of the period is that it is eminently and fully human.

At this remove it is hard to say what kind of accident or design the scratches on Lower Palaeolithic rocks and bones might be, though this does not weaken the convictions of the scholars who study them. Robert Bednarik believes they were animated by 'concepts'; Randall White scoffs at the idea that they show 'any more patterning than is exhibited by my kitchen cutting board'.[3]

Some may be natural marks, the signatures of fractures or blood vessels; others may indeed be scratches produced in the course of practical activities such as butchery or cutting hide. Even if they were made intentionally, they were no more sophisticated than the drawings of three-year-old children, or modern domesticated chimpanzees. The fan-shaped patterns are actually quite similar to ones drawn by chimpanzees under the supervision of Desmond Morris in the early 1960s.[4]

Children and chimpanzees do not make their own crayons or paper. They have a potential to draw, but it is only realized within a culture made by adult humans. A few chimpanzees have learned to communicate using graphic symbol systems devised by their human mentors. Whether their cousins in the wild are capable of devising their own symbols is unclear, though intriguing reports suggest that bonobos may mark trails with flattened plants and branches stuck in the ground.[5]

If you were parachuted into the West African forest, without a briefing on bonobo habits, and you stumbled upon a branch stuck in the ground, you would probably assume that someone had put it there. 'Someone' implies people; 'something' is only appropriate for monsters. Neither seems right for bonobos. They are certainly not monsters, and they are not people either. Our own symbol systems are poorly suited to representing modern minds that are not human, or ancient minds that are human but different from ours. Was the individual who inscribed the Tata nummulite someone? Robert Bednarik thinks so, referring to the engraver as a 'person', an 'artist' who was not satisfied merely to possess objects with aesthetic qualities, but 'improved them, commented upon them'.

It was a one-line comment, though, and not a uniquely human one either. Desmond Morris observed that a chimpanzee named Alpha and one called Joni showed 'a strong inclination to cross boldly-drawn lines or bars at right angles'.[6] Alexander Marshack has described a more complex Middle Palaeolithic design, a pattern of nested semi-circles and vertical lines on a piece of flint, which, like the Berekhat Ram figurine, was found on the Golan Heights. The site of its discovery, near Quneitra (in 'Demilitarized Zone A'), has been dated at 54,000 years before the present; Marshack hailed the carving as 'the earliest known depictive image'.[7]

Marshack's use of the microscope and his belief in detail have created a powerful body of work which discovers system in what many lay people, and scholars, would dismiss as meaningless doodles. His showpiece analysis identifies a thousand marks arranged in rows on a flat piece of bone, from the Grotte du Taï in eastern France, as an Upper Palaeolithic lunar calendar.[8] He has asserted the existence of ordered representations in deep prehistory, in the face of a tendency among archaeologists to regard them as oddities that distract from the real art of the Upper Palaeolithic. They are fascinating precisely because they are not part of a recognizable culture.

There will always be pleasure in devising explanations for the 'Franco-Cantabrian' cave art that includes Lascaux and Altamira, as prevailing accounts move from 'hunting magic', to Leroi-Gourhan's structural model based on sexual symbolism, and thence to shamanic hallucinations, which are currently enjoying something of a vogue. David Lewis-Williams is the leading advocate of the 'entoptic' interpretation of rock art, which turns upon the apparent universality of certain images which do not correspond to something in the external environment. At the lowest level are phosphenes, the points of light involved in 'seeing stars'. (Robert Bednarik has a rival account of rock art based upon these.) Then come zigzags, spirals, lattices, nested curves, parallel lines and the other building blocks of hallucination.[9] They can be induced by migraine attacks, hallucinogenic drugs, or sometimes just by pressing the eyeball.

Nor are they confined to the mind's eye. Oliver Sacks and Ronald K. Siegel observe that patterns like these are frequently produced by complex systems. Like Brian Goodwin, they follow the tradition of the structural biologist D'Arcy Thompson and his book *On Growth and Form*, which was published in 1942. Sacks and Siegel suggest that these patterns are fundamentals of nature. Although they do not represent anything outside the mind, they sometimes resemble objects encountered in the real world. 'A migraineur . . . cannot open *Growth and Form*, cannot see its pictures of radiolaria and heliozoa, of pinecones and sunflowers, or spirals, lattices, tunnels, and radial symmetries, without a startled cry of recognition.'[10] Patterns as prevalent as these were surely seen in the eyes of ancient minds as well as modern ones. The cry of recognition might have been heard in the Cure valley 30,000 years ago, as a Neanderthal eye lit upon a fossil of a spiral shell or a spherical madrepore.

Lewis-Williams organizes altered states into progressive stages. Entoptic patterns form the overture; then, as the hallucinated condition deepens, the mind attempts to construe these patterns as familiar objects. In the deepest stage, the mind concentrates on iconic hallucinations of animals, people or events; the entoptic patterns are relegated to peripheral decoration. He developed his theories by studying the rock art of the South African San people, and then went on to collaborate with the French prehistorian Jean Clottes in proposing them as a means to understand the Upper

Palaeolithic cave art of France. Instead of Leroi-Gourhan's 'mythograms', Clottes and Lewis-Williams suggest that insight will come from seeing the French images through prisms of shamanism and neuropsychological universals.[11]

At the level of figures, the French cave paintings are immediately comprehensible. Writing in the journal *Perception*, John Halverson observes that 'Paleolithic [sic] art is thoroughly grounded on the principles of formal simplification and accentuation of salient features'.[12] Figures are delineated by outlines, and distinguished by features peculiar to the type of creature that they depict. Features common to a number of types, such as legs, may be vestigial or absent. Any modern adult and most small children will recognize a Chauvet rhinoceros as a rhinoceros, because it has rhino horns. We can be fully confident that it represents a rhinoceros, and that it signifies other things besides. Although the nature of those things is inaccessible to us now, cave art is intuitively recognizable as the work of modern humans, and as part of a symbolic culture. If such a culture were to exist today, any child raised in it would be able to acquire it, given normal abilities but regardless of ethnic origin, just as any such child would be capable of acquiring any actually existing culture in which it was raised. Although modern children are capable of much more than their Lower Palaeolithic forebears, the radical difference between them would keep them apart.

Whether they would be separated in the same way from children of the Middle Palaeolithic is an interesting question. During the period in which the Quneitra stone was carved, 54,000 years ago, both modern humans and Neanderthals lived in the Levant. They used similar tools, classified today as Mousterian, and there is no particularly strong reason to assume that an individual of one type or the other made the marks – other than the circular argument that because it looks like a graphic design, it is probably the work of a modern human.

The distinctions between Neanderthals and moderns are

altogether more blurred in the eastern parts of the Neanderthals' range, where their distinguishing physical characteristics were less accentuated. Even in the western homelands of the 'classic' Neanderthals, with their heavy brow ridges and bodies built to match, there is now strong evidence that some of them ended up with definite cultural similarities to the people who replaced them. As well as the fossil collections at Arcy, three dozen beads and pendants have been recovered from Châtelperronian layers. They are made from animal teeth, ivory, bones, fossils, and a stalactite. By the use of computed tomography scanning, scientists have identified a human bone found in one of the Châtelperronian levels as that of a Neanderthal child, who died at the age of one. Although this is only the second Neanderthal to be found among Châtelperronian artefacts, it is strong evidence that the industry was part of the Neanderthal way of life.

How it came to be there is another matter. A number of scientists doubt that it was the work of Neanderthals alone. Perhaps they acquired their beads by trade with moderns, or learned how to make them from their more capable fellow hominids, or just found them lying on the ground. Neanderthal sympathizers, a vocal current in palaeoanthropology, consider this patronizing at best. It is easy to hear the echoes of European assumptions about racial supremacy, and to wonder if this is not a projection of the scene from the old textbooks, in which Africans trade each other for the Europeans' beads. Vivid but dubious claims about Neanderthal cannibalism also seem to belong to this picture. No wonder some of the sympathizers founded a Neandertal [sic] Anti-Defamation League.

Their cause receives support from a review of the Arcy-sur-Cure evidence by Francesco d'Errico and colleagues.[13] Against the idea that Châtelperronian artefacts were acquired from moderns, the authors point to evidence of bone-working which indicates that some of the objects were made on site. They also consider that the objects differ in too many respects from

Aurignacian artefacts, representing the first clearly Upper Palaeolithic industry, to be derived from them. Their vision is of two distinct cultures side by side, Neanderthal and anatomically modern. There may have been exchange between them, including that of some Aurignacian beads found in the top Châtelperronian level, but this was interaction rather than acculturation, which in their book is a euphemism for swamping by an alien culture. All in all, the scholars conclude that 'the archaeological record of the Middle-to-Upper-Paleolithic [sic] transition in Western Europe provides no material support for the theories developed to accommodate the Arcy record to the notion of "Neanderthal inferiority"'. Although a number of their peers take issue with aspects of their argument in comments that follow the paper, several are also at pains to deny that they believe in the biological superiority of modern humans. For some peoples, after all, the Stone Age lasted until the twentieth century. The technological gap between them and the people who descended into their lives in aeroplanes is far greater than that between Mousterian and Aurignacian stone tools; but now some of them fly planes themselves.

Robert Foley and Marta Lahr have argued that the more meaningful line of division in human prehistory is that between the Lower and Middle phases. The earliest remains of modern humans are associated with Middle phase tools; the Upper Palaeolithic is merely a regional phenomenon, confined to Eurasia, the Levant and north Africa. As far as human evolution is concerned, the Upper Palaeolithic is less significant than the appearance of Middle Stone Age tools in Africa around 250,000 years ago. These, Foley and Lahr suggest, are the signs that something big has happened to the mind, not the cultural explosion that begins around 40,000 years ago.[14] Alexander Marshack points in the same direction when he suggests that 'the nature, content and variability of early Upper Palaeolithic symboling in Europe suggest not a beginning or origin but rather the products of a long preparation'.[15]

If it were otherwise, humankind as we know it would have had to be created by the genetic equivalent of a thunderbolt from the gods. Suddenly a mutation transforms the mind, like the monolith which implants the knowledge of evil into the australopithecines in the film *2001*. As far as we can tell, human anatomy has not changed over the past 100,000 years in any way that suggests a sudden leap in mental capacity. Yet a spectacular flowering occurs in human behaviour in the period between 50,000 and 30,000 years ago. Any fateful mutation would have had to have been an upgrade within the existing case, since the skulls remain the same before and after the symbolic revolution erupts. Or it could have exerted its effect upon the peripheral soft tissues, as Jared Diamond has suggested. He thinks the 'Great Leap Forwards', as he calls it, was made possible by the fine tuning of the vocal tract.[16]

A mutation acting inside the skull is possible too. It is implicit in Steven Mithen's model of cognitive fluidity, in which the modern condition is attained by knocking through the last of the walls between the mind's various modules. Such a change, requiring only that the brain's networks grow denser, would leave no impression on the fossil record. It might not be all that expensive, metabolically speaking. But as a concept, it's a bit too easy. There is something unsatisfactory about invoking a gene when there is no biological evidence of its effects. When Jared Diamond refers to the Great Leap Forwards as 'that magic moment of evolution around 40,000 years ago, when we suddenly became human', he inadvertently puts his finger on what is wrong with the Great Leap Gene: it smacks of magic. And this is the origin of human culture we are talking about, after all, not the origin of walking upright. Culture is capable of reproducing itself: why assume that a gene was needed to switch it on in the first place?

In their assault on the idea of 'Neanderthal acculturation', Francesco d'Errico and his colleagues also point out that biological determinism is too easy: 'If everything is in the wiring of the brain, why should we bother with artifacts, settlement patterns, or faunal exploitation strategies?' They indicate a different way of imagining human fortunes in the early stages of symbolic culture. 'Once this reductionist fog is cleared, a much richer world appears – a world in which ecological diversity and historical contingency created a variety of biological and cultural actors.' Different populations developed unevenly, according to circumstances rather than innate capacities. Some of the French Neanderthals created the Châtelperronian industry; their Spanish counterparts continued with the Mousterian industry that is most typically associated with Neanderthals. Even if their skulls were a different shape from those of the anatomically modern people who replaced them, their brains were as big, if not bigger. And even if they did learn their Châtelperronian tricks from the incomers, or acquired their beads by trade, the simple fact that they incorporated them into their way of life shows that they had the mental capacity to appreciate these facets of the modern human condition. The Neanderthals' beads suggest that it might be a good idea to look outside the genome when trying to identify what it was that gave our ancestors their edge.

Beads, and other forms of body adornment, indicate a sense of personal identity. There is also strong evidence that Neanderthals buried their dead, but opinion is divided as to whether this had symbolic significance. Remains of one Neanderthal child, found at La Ferrassie in France, were covered by a large stone into which cupules had been dug. These holes seemed to be distributed more or less at random, prompting Alexander Marshack to suggest that they might have been marks of ritual, rather than symbols that conveyed meaning through pattern. Marshack imagined the members of the funeral party each carving out a hole to show that they had taken part. There are backward echoes here, of the Jewish custom of placing stones upon a grave.

Other prehistorians doubt that Neanderthals had a spiritual dimension. Neanderthals do seem to have put their dead in the

earth, to coin the phrase that William Golding gives to the dying
Neanderthal in his novel *The Inheritors*, but the sceptics suspect
this was simply a convenient means of disposal. They point to the
bareness of the graves. Sometimes objects such as animal bones or
antlers have been found near the buried remains, but there is no
really unarguable evidence of grave goods or other symbolic elements. Golding's Neanderthals did not need artefacts to sustain
their belief that the goddess Oa had 'brought forth the earth from
her belly', and that she had taken their companion back there.[17]
William Golding made free use of his novelistic licence to create
an entrancing vision of a radically unfamiliar human mind, complete with telepathic powers. Archaeologists are bound to the
material record, however, and the evidence that the real
Neanderthals had a sense of self is much stronger than the evidence that they had a sense of soul. Perhaps that was the secret of
the inheritors' success.

Many things were needed for the development of symbolism,
language and human culture in its familiar form. Before adult
humans can speak fast and utter a wide range of sounds, for example, the larynx has to move downwards. This makes it impossible
to eat and breathe at the same time, as apes and children can, and
early hominids almost certainly could. The benefits of speech
must have been immense, since they would have had to outweigh
the risk of choking that the shift introduced. Another requirement was the ability of listeners to process speech in chunks,
parsing the stream of sound into syllables and words. The extent
to which Neanderthal anatomy had made the necessary adjust-
ments is uncertain and hotly controversial. But no amount of
engineering developments could ever be enough on its own. To
become human as we know it, human beings needed trust.

3

Ochre comes in shades of yellow, orange, red and brown; the core of it is the iron red of ferric oxide. Together with manganese dioxide, which is densely black, it offered Palaeolithic hominids a palette covering the spectrum of fire, from flame to charcoal.

Up to about 110,000 years ago, they only dabbled occasionally. Then, in southern Africa, ochre seems to have coloured their whole lives. It is present in 'copious' quantities at every cave and rock shelter that contains relics of occupation from this period. This was a monochrome explosion, based almost entirely on red ochre, and particularly upon strong reds, rather than orange or brown shades. The collectors of pigment made little or no use of the deposits of manganese ore and magnetite, a black iron oxide, which were available in some areas. According to Ian Watts of University College, London, who has made the study of prehistoric ochre his own, 99.5 per cent of all known African Middle Stone Age pigment is iron oxide, and 94 per cent contains a red streak.[18]

The sudden red dawn, in a late stage of the Middle Stone Age, was an isolated one. In Europe, the pigment least infrequently used between 70,000 and 35,000 years ago was manganese dioxide. Just one Mousterian object in the published literature seems to share the character of the African ochre phenomenon. Described as a 'plaque' and estimated to be 100,000 years old, it is made from part of a mammoth molar tooth, and coated in red ochre. Like that other flash of Mousterian genius, the nummulite fossil with a line inscribed at right angles to a natural fracture, the plaque was found at Tata in Hungary.[19] Otherwise, ochre features hardly at all before the Châtelperronian, at which point Neanderthals suddenly adopted it. Nearly 20kg of the stuff have been recovered from one Châtelperronian level at Arcy-sur-Cure, the major Neanderthal ochre site.

Several of the southern African ochre specimens have been modified in ways that do not appear to have any practical purpose. They have notches cut out of them, holes drilled into them, and lines scored upon them; some of the latter form patterns, such as parallel lines or triangles. These were not just tools for making decoration, but objects of decoration themselves. They add to the unmistakable scent of symbolism that hangs over Palaeolithic ochre.

4

Suppose that a party of Boxgrove hominids has made its way down the gully that cuts through the cliff, and is moving towards a waterhole. Glancing back, a juvenile notices that a wolf is standing by the stream, on the path they have just taken. To warn his companions, he cries out and points. The wolf is gone. The young hominid now launches into an energetic performance, gesticulating, posturing, executing steps like a modern human dancer, repeating a distinctively shaped sound over and over. He uses his whole body and a wide circle of space around it, as well as his voice.

His companions seem impatient at first, some turning away as if to resume the trek. This rouses him to still greater efforts, and eventually he commands their attention. Now they are all facing the cliff. The juvenile falls silent and squats down, as do some of the others. They gaze at where the wolf had been, as the sun passes by overhead.

Lower Palaeolithic jokes probably didn't have punchlines either.

Aesop's fable of the boy who cried 'wolf' is leaden in Acheulean translation because of the effort needed to communicate without a symbolic culture. If the wolf had stayed where it was, the warning would have been simple. The juvenile would have only to signal alarm and the rough location of the cause for it. He might be able to use a specific sound to signify 'predator', or even 'wolf'. His companions would look in the direction he indicated, and verify for themselves that there was an object corresponding to his signal at the spot.

Unfortunately, the chances that they would be able to obtain this confirmation are reduced by the noise and commotion of the alarm signal itself. Without a symbolic order, everybody has to shout. Every signal has to carry its own conviction, and conviction is indicated by cost. Cheap signals are easy to fake, and so cannot be trusted.

If the wolf hears the alarm and makes itself scarce, the juvenile is immediately faced with a far more complex problem of communication. He has to convey the message that an important object was in a certain place, and justify the call on his companions' attention. The more information he can convey, the greater his chances of convincing the others will be. He is now under pressure to fill in the picture with more detailed information about the location and nature of the object, in order to demonstrate its significance. A wolf on the trail back to where they spend the night is more important than a wolf in the same general direction, but on the clifftop above. A group of horses would also be worth noting, and so would a thundercloud, but each would require a different response.

In a situation like this, the temptation is to exaggerate, or to lie. The audience has to decide whether to believe the performer. Like the Greeks to whom the shepherd boy cried wolf, the Acheuleans can draw on past experience to assess his trustworthiness. In the main, though, they have to gauge a message about an invisible object by the effort the signaller puts into it, and by whether they can verify it themselves. Both these measures are likely to require considerable effort on the audience's part. They cannot simply listen to the signaller's words while they carry on with what they were doing before. They have to stop, look and listen, in order to assess both the cost and the meaning of the signal. Once they are convinced that the juvenile is likely to be telling the truth, even if they are not quite sure what that is, they then have to see it for themselves. That requires staying put and staring, until the wolf reappears or they have had enough.

If we could see them now, they would probably look pretty stupid, with their slapstick antics and their solemn stares into the distance. It's true that they might well have found communication easier if their mental capacities had been greater, but that was not their fundamental problem.

Nor would it have been all right if they could have talked properly. The question of becoming human raises questions of self-consciousness, culture and language. Many scientists put language first, as the most impressively exclusive human trait, and many are inclined to treat it as a problem of machinery. This approach has been strikingly productive, leading to disputes as heated as any in the field of human origins.

Even the earliest hominids have been called in evidence. Most of what little information can be gleaned about hominid brain structure comes from the impressions that brains have left on the inside of fossil skulls. Dean Falk and Phillip Tobias believe that they can discern the mark of a region called Broca's area in the brain of very early *Homo*, but not in australopithecines. Broca's area has been associated with speech since 1861, when the anatomist and anthropologist Paul Broca identified it as the site of damage to the brain of a man known to hospital staff as 'Tan' – the only sound the patient could utter. It is hard to pin down, though, and monkeys have similar structures. Even if the convolution was one of the features that distinguished hominines from australopithecines, it may have arisen to perform some other function, long before it was recruited to serve language. Stephen Jay Gould and Elizabeth Vrba coined the term 'exaptation' to describe instances in which selection appropriates a structure for a new purpose. Exaptation is considered to be standard procedure in evolution, but its suggested effects can be radical. One explanation for the evolution of insect wings, for example, is that they originated as fins for dispersing heat.

At the other end of the hominid timeline, the most heated controversies are located not in the brain, but just underneath it. The question at issue is whether any hominids other than anatomically modern ones had structures for sound production sophisticated enough for efficient speech. The vocal tracts of modern adult humans are distinguished from those of infants and other mammals by the low position of the larynx. This increases the volume of air available for modification during speech, and directs the outward impulses of air towards the mouth rather than the nose. These alterations make for clearer and more varied sounds, at the price of an increased risk of choking. In the words of Jeffrey Laitman, a pioneer of hominid vocal tract reconstruction, the effect of lowering the larynx is like turning a bugle into a trumpet.[20]

The position of the larynx in ancient hominids can be inferred from the shape of the base of the skull. Those of some Neanderthals are fairly flat, suggesting a high larynx, but those of some of their likely ancestors are curved, indicating a larynx in a characteristically modern low position. According to Laitman's colleague, Philip Lieberman, *Homo erectus* did not have a larynx low enough to support speech. The necessary vocal apparatus would therefore have emerged somewhere in the grey area between *erectus* and *sapiens*, during the past half million years.

At least one element of the system does appear to have been in place by this time, judging by another feature of the base of the skull. The hypoglossal canal is the duct through which run the nerves from the brain to the muscles of the tongue. Three researchers from Duke University, in North Carolina, have compared the width of the canals in contemporary apes and extinct hominids. Chimpanzees, australopithecines, and a specimen assigned to *Homo habilis*, had canals of similar bore. Those of modern humans are twice as wide; most of the difference between apes and humans remains after corrections have been made for the different sizes of the primates' mouths. The team reasoned that the canals might be wider in order to accommodate more nerve fibres, necessary for controlling complex tongue movements (though their first study did not exclude the possibility that blood vessels took up the extra space). Since eating and drinking make basically similar demands on the tongues of apes and humans, speech is the obvious candidate to explain the increase in bandwidth.

If the difference relates to language, it is not surprising that very early hominids resemble apes rather than humans. This is a rare example of a finding that ruffles few feathers in the fractious domain of language origins, in which one camp thinks that language developed quietly and gradually throughout the course of hominine evolution, while the other maintains that language appeared late and all at once, immediately becoming the wellspring of the cultural deluge which marks the Upper Palaeolithic. The Duke researchers measured the hypoglossal canals of two skulls from the grey area of archaic *sapiens*, or *heidelbergensis*, or *rhodesiensis*. In these specimens, 200,000 or 300,000 years old, the widths of the canals were within the modern human range. So were those of two Neanderthals and an early modern human. Matt Cartmill, one of the researchers, suggested that Neanderthals 'had tongues as nimble as yours'.[21]

They also had a bone in the throat like ours, called the hyoid. When a fossil Neanderthal hyoid bone was discovered at a site on Mount Carmel, in Israel, it was hailed as a token of Neanderthal speech. Critics objected that its position in the throat was unknown.[22] Since much of the vocal apparatus is made of soft tissue, the course of its development will always be uncertain.

Chris Stringer and Clive Gamble point out that once the shift towards language had taken place in the Neanderthals' ancestors, its reversal would be unlikely. The anatomical evidence may remain equivocal, but if the hypoglossal canals were indeed filled with a fat bundle of nerves leading from the undoubtedly large Neanderthal brain, it seems reasonable to suppose that Neanderthals had some sort of capacity for speech. If their larynxes remained short, they might have spoken a language with a limited range of vowel sounds. That would not have been such a terrible handicap, though. After all, the English upper classes managed without the vowel 'a' for half the twentieth century.

There may be a threshold of performance below which language as we know it cannot be sustained. People can process information about three times faster when it comes as speech than when it takes any other form. More than twenty units of information, each roughly corresponding to a letter, can be uttered and comprehended in a second. In any other modality, fewer than ten units of information can be processed in a second. If speech was as slow as that, a typical sentence might exceed the capacity of short-term memory. With a slow communication system, hominids would be limited to simple utterances. Their conversation might not have been sophisticated enough to stimulate the development of a comprehensive language faculty. The level of such a threshold is a matter of guesswork, but it seems reasonable to assume that the Neanderthals were well above it. They are simply too close to us.

Although the Neanderthals are the co-stars of human evolution, with their hulking charms and noble savagery, older varieties also raise interesting questions about speech. If they had nimble tongues as well, what were they doing with them? One possibility is that hominids, such as those of Boxgrove, could articulate a large range of sounds, but had not developed the capacity to organize them with syntax. They might have had words, but no language.

A lack of grammar was not the reason they were dancing and gesticulating in the imagined scene at the beginning of this section. Mime and gesture were not needed as a makeshift scaffold to give communication a structure. They were needed to overcome a lack of trust. Their powers of persuasion came partly from the sheer effort involved, and partly from the distribution of signals across modalities. As all but the most perfunctory messages were likely to require the use of facial expression, gesture and bodily movement, as well as sounds, there simply were no short cuts by which they could be delivered cheaply.

The idea of mime derives from the psychologist Merlin Donald, who has proposed that hominids passed through a stage in which they based their communication upon it. This was the first great human leap forward, establishing a mode of thought more fundamental than language, and independent of it. 'Mimetic action,' Donald writes, 'is basically a talent for using the whole body as a communication device.'[23] What made it possible, he argues, was a revolution in which hominids took control over their bodies.

Thanks to the work of Dorothy Cheney and Robert Seyfarth, the alarm calls of Kenyan vervet monkeys are among the most celebrated sounds in primatology.[24] Vervet alarm calls sort the monkeys' many predators into four classes. There is one call for large cats, either leopards or cheetahs; one for eagles; one for snakes, either mambas or pythons; and one for primates, baboon or human. On hearing an 'eagle' alarm, vervets look up at the sky; on hearing a 'cat' warning, they scramble up trees.

Marc Hauser, a theorist of animal communication, has observed a degree of flexibility in the system. Driving through the Amboseli National Park, where Cheney and Seyfarth have conducted their studies, Hauser heard vervets giving 'cat' alarms.

These calls sounded wrong, though. They were slower than usual, as if 'the batteries of a tape recorder were run down during play-back'. The cause of the alarm turned out to be a lion, a cat so large that hunting vervets would not be worth its while. 'Slowing the tape' was an appropriate response, but there is not much more that can be done with a tape or a vervet call. The monkeys' alarms are reliable because their only legitimate meaning is that a vervet has just seen a predator. Either the calls will be verified immedi-ately, or not, in which case they will be disregarded. The monkeys who send the messages cannot elaborate them – unlike the Acheulean juvenile who cried 'wolf' – so it only takes a glance for the other monkeys to check their veracity. While a system like this deals efficiently with a present and visible threat, it cannot cope at all with even the recent past.

Though apes have much more complex cognitive faculties than monkeys, they remain generally unable to exert deliberate control over their signals. Sometimes chimpanzees manage to stifle their cries, but only with difficulty. If they see something exciting, such as food, they generally cannot help but let the cat out of the bag. They do not cry 'food', though. Despite their much greater cog-nitive capacities, chimpanzees do not attach particular meanings to particular calls.

The reason may be that they are too clever for their own good. A boy in the playground is being threatened by a larger boy. He points over the larger one's shoulder and cries 'Behind you!'. 'That's the oldest trick in the book,' retorts the aggressor. Judging by an incident recorded by Richard Byrne and Andrew Whiten, the trick may be very, very old indeed. A sub-adult baboon was harassing a younger animal, which brought a party of adults to its aid by screaming. When the harasser saw the adults coming over the hill, he jumped up on his hindlegs and stared across the valley into the distance, as if he had seen a predator. His challengers stopped and stared in the same direction, instead of pursuing their attack.

When they surveyed the primatological literature, Whiten and Byrne found that deception was part of life in all monkey and ape families. It typically took the form displayed by the adolescent baboon, boys in playgrounds and slapstick comedians, of attempts to manipulate the attention of others.[25] The reason the trick is still part of human culture, millions of years after it first evolved, is that it can still work if the cheat can produce a convincing impression of an involuntary response. A look of alarm and a spasm of muscular recoil may trigger an involuntary response on the part of the victim, that makes him wonder for a fraction of a moment whether there really is a monster behind him; and a fraction of a moment's uncertainty is all a clever cheat needs.

With deception so deeply ingrained in primate life, it is not surprising that vervet monkeys are unusual in having signals with specific meanings. It is probably a condition of their existence that these signals are tuned to the delivery of messages which are likely to benefit all the individuals who hear them. Chimpanzees lack predators, apart from humans, and so would not have common interests in alarm calls of the vervet type. Their overwhelming concern is with each other, and their intelligence is needed to keep up with greatly elaborated Machiavellian challenges of the type demonstrated by the cunning baboon. Their intelligence resides largely in their Machiavellian capacities, but these very faculties prevent them from developing anything like a lexicon of signals. As Chris Knight puts it, they are too clever for words.

The fact that several chimpanzees have proved able to use symbols devised by humans, amounting to visual words, only highlights the question of how humans, alone among living primates, have established the basis of trust for symbolic communication. This was surely the most fundamental of transformations in the process of becoming human as we know it. But it does not imply that before trust hominids lacked language entirely. Whatever forms their societies took, all would be different to greater or lesser degrees from those of chimpanzees. The marked differences between bonobo and common chimpanzee social relations illustrate how much a relatively minor shift in the balance of power can influence the quality of life. As hominids acquired the capacity for more sophisticated forms of behaviour, the benefits of co-operation may have increased. By the later Acheulean period, hominines appear to have anatomical adaptations for speech. The debates over the extent of early language capacities are baroque, highly strung, and best avoided if possible. This account sticks to the Darwinian point that, whether hominines had words, syntax or both, they were limited by the extent to which cheating could manipulate their communication systems.

In the case of the Acheulean juvenile who cried wolf, the group would have been presented with a message about something on which it would not be particularly easy to agree a meaning. A wolf would not have the same implications for middlebrow hominids as a leopard for vervet monkeys. For vervets, a leopard is a predator, posing a straightforward threat. Although it would be going too far to claim that hominines had no predators, the threat they faced from wolves and other carnivores would have been more equivocal. Wolves would have learned to respect the defensive capabilities of intelligent hominines who used weapons and threw missiles. They would have been opportunistic rather than regular predators, snatching juveniles when hungry enough to make the effort worth while, or when the odds happened to favour them. Hominines and wolves would often, however, have been competitors for the same meat. So wolves would always have been of interest to hominines, but the appropriate response to a sighting would vary greatly according to circumstances. Among vervets, the alarm system is completed by the response rather than whatever the monkeys understand by the signal. In that sense, the call triggered by a leopard means 'climb a tree' rather than 'leopard'. Without such an urgent and unequivocal message, hominines would be unlikely to develop a sign for 'wolf'.

Whether their brains are large or small, then, primates are confined to the same frame of reference. If they can see an object, they can and often must refer to it; if they cannot perceive it, they cannot refer to it. Primate communication systems will remain earthbound until their signals break free from the concrete objects they represent. According to Merlin Donald, this happened when hominids developed a mimetic faculty, which required them to have access to their memories at will. According to Chris Knight, on whose fundamental insight about trust this discussion is based, the next stage in the liberation of signals required the invention of deities. What the Acheulean juvenile needed was a god to swear by.

5

In his book *Blood Relations*, published in 1991, Chris Knight recalls how he came to appreciate the powers of sociobiology. He had by then spent more than twenty years following twin tracks of political activism, rooted in Marx, and studies in cultural anthropology, based on Lévi-Strauss. Precisely because of its radical calculating individualism, Knight saw in sociobiology the same kind of revolutionary power that Marx had seen in capital. In an incandescent passage of the *Communist Manifesto*, Marx and Engels hailed the achievements of the bourgeoisie, which had destroyed feudalism and 'drowned the most heavenly ecstasies of religious fervour, of chivalrous enthusiasm, of philistine sentimentalism, in the icy water of egotistical calculation'. Sociobiology's ruthless cost-benefit calculations, Knight realized, had a similarly devastating impact upon 'well-meaning' theories of selection for the good of the group. Its achievements, he wrote, 'are the corrosive acid which eats away at all illusions, all cosy assumptions about "the welfare of the community" or the "brotherhood of man", all unexamined prejudices about how "natural" it is for humans to co-operate with one another for the good of all'.[26]

A few years later, Daniel Dennett hit upon the same metaphor to convey the power of 'Darwin's dangerous idea'. He compared the idea of evolution by natural selection to universal acid, an imaginary substance with which he and his friends would play as schoolchildren. Universal acid dissolves everything, and so cannot be contained. Like it, Darwin's idea 'eats through just about every traditional concept, and leaves in its wake a revolutionized world-view, with most of the old landmarks still visible, but transformed in fundamental ways'.[27]

Dennett affirms that culture is grist like anything else to the mill of Darwinian selection. His thoughts are mostly of memes, the loose equivalent of genes in the world of ideas, and how Darwinian processes may underlie their spread. Knight's focus is on how culture could have been created by animals that were subject to Darwinian processes in the conventional biological sense. Dennett speaks of 'good tricks' that evolutionary design will often employ. Knight and his colleagues have constructed a trick of extraordinary cunning, which they propose as the means by which Darwinian animals reconciled their interests enough to make possible language and symbolic culture.

Blood Relations is an extraordinary work, in which imaginary creatures and magical events are orchestrated on a global scale, from Australia to Amazonia, into a single vision of how humans created humanity. Speaking at an event billed as a 'Great Sociobiology Debate', for the motion that 'Darwinism can explain the origin of culture', Knight declared that it was not the use of tools by chimpanzees that needed explaining, but how a species comes to be able to distinguish between water and holy water. Nobody is more impressed by the power of capitalism than a Marxist, and perhaps likewise it takes a Catholic upbringing to realize just how far Darwinism can go.

Though Knight does tend to resemble a shaman with a spreadsheet, he is not concocting some syncretic religious brew of Darwinism and tribal initiation rites. He is every bit as materialist as Dennett or Dawkins – ultra-Darwinian, in Stephen Jay Gould's terms – but unlike them, he has an intuitive understanding of the sacred. The trick here is to retain one's sense of magic after one stops believing in it. *Blood Relations* appreciated the importance of sacred ritual, and of sociobiology, the better for being able to stand outside them. Writing under the influence of *Primate Visions*, Donna Haraway's feminist interpretation of primatology, Knight felt able to refer to his own narrative as myth, and free to bring his own props to the sociobiology show. 'If you could have calculating, maximizing capitalists operating in human origins narratives, why could you not *also* have militant trade unionists?' he asked. 'If you could have profits and dividends, why not also industrial action, pay bargaining and strikes?' Culture, he proposed, was the settlement that followed the world's first strike.

Women's reproductive cycles have several distinctive features that add up to a unique combination among primates. There is no signal to indicate that ovulation has taken place, but halfway between ovulations, bleeding occurs. In contrast to females of other species, women may mate at any stage in their cycles. And the mean length of the cycle, 29.5 days, is the same as that of the Moon's cycle from full to dark.

Although imperceptible ovulation is a distinctive feature of human sexual physiology, it is not a unique one. It seems to have arisen a number of times in different primate lineages. Birgitta Sillén-Tullberg and Anders Møller surveyed the mating systems used by species in which ovulation is concealed, to weigh the balance between two competing explanations for this concealment.[28] One school of thought has argued that concealed ovulation promotes monogamy by inducing a male to remain around a particular female for longer, in order to increase the chances that mating will take place during a phase when an egg is present for fertilization. It is thus a device for increasing a father's confidence about who his offspring are.

The other school argues just the opposite: that concealed ovulation is a Machiavellian tactic to confuse the issue of paternity. If a male is uncertain whether or not he is the father of a juvenile, he is less likely to harm it. Concealed ovulation is thus a device for making males behave better. Sillén-Tullberg and Møller concluded that it might have evolved once in a monogamous species, or not at all, whereas it seems to have evolved between eight and eleven times in species with non-monogamous systems. The evidence therefore supports the Machiavellian hypothesis, rather than the monogamous one. On the other hand, the researchers found that of the seven times that monogamy itself evolved, four to six of these events took place in the absence of ovulatory signs. So once concealed ovulation has evolved, even if it has done so as an adaptation to a non-monogamous mating system, it is then conducive to the evolution of monogamy.

Under mating arrangements that are far from monogamous, such as a harem system in which a single male exercises a reproductive monopoly, it would suit the male for the females' reproductive cycles to be unconnected to each other. If all the females reached the fertile phases of their cycles at the same time, the male would probably succeed in impregnating far fewer of them than if their windows of reproductive opportunity were randomly distributed. Conversely, females could thwart reproductive monopoly by synchronizing their cycles. A trend towards female synchrony would be attractive for males even when more than one of them had already achieved reproductive success. If males were staying with single mates, they would be competing less with each other.

In 1979, Nancy Knowlton published a paper which pointed out that cycle synchrony could be a strategy for encouraging males to invest in their offspring. If all the females in a group have synchronized cycles, there is little point for a male in abandoning a mate once her fertile period has finished, since none of the other females will be fertile either. He will do better to stay with her and invest his energies in the welfare of their offspring, and he will be the more inclined to do so because of the increased likelihood that he really is the father.[29]

Applying the logic to the chimpanzee-like primates he took to be at the root of the hominid lineage, Paul Turke then devised a scenario for the evolution of distinctively hominid female reproductive cycles.[30] If females signalled their fertility conspicuously, as female chimpanzees do with the genital swellings which accompany oestrus, males of high rank would mate with the most conspicuously fertile females. Lowlier males would mate with females with lesser signs of oestrus. Although the lowly and the less obviously fertile would not be the mates of first choice, both would have attractions of their own as a result. A male whose mate had muted signs of oestrus would face less interference from other males. A lowly male would be more likely to stay with a female, having less chance of success elsewhere.

Turke suggested that a female whose oestrous signs were lower key but longer lasting than those of others could extract higher levels of investment from a mate than her rivals. She would attract a low-ranking mate, who would be less likely to leave her to pursue other opportunities. The longer her oestrous period lasted, the longer he would be likely to stay, driving a selective trend towards what is questionably described as 'continuous receptivity'. Synchronizing with other females would complete a package adapted to securing male investment, by promoting tendencies towards monogamy. Turke's model encapsulates what sociobiology did for females. A radically individualist paradigm, based on conflicts of interests, placed the interests of females at the centre of the stage; and showed how individual interests could be reconciled into collective ones.

Ovulation in women is undoubtedly concealed, even from women themselves, as the difficulties of encouraging or preventing conception without artificial means affirm. Menstrual synchrony is another matter. In humans, rather than rats or golden hamsters, it has proved an elusive phenomenon. Martha McClintock first described it in 1971, but it has still to be accepted as real by many scientists.[31] There are two main reasons for the enduring scepticism. One is that while many studies have detected synchrony, many have not. A review published by Leonard and Aron Weller in 1993 gave a tally of sixteen studies reporting synchrony, and eleven which failed to find it.[32]

Even in studies with positive findings, it is impossible to say why some women synchronize their cycles and others do not. In a survey the Wellers conducted on lesbian couples, ten of the couples did not synchronize cycles and ten did. The only thing that the researchers could find which appeared to encourage synchrony was eating together.[33]

Roommates, friends, workmates, mothers and daughters have also been studied. Thanks to a Bedouin nurse who conducted interviews among women of her village in northern Israel, the Wellers were able to obtain information about synchrony under conditions they regarded as optimal: the Bedouin women lived together for many years, were segregated from men, were extremely unlikely to have sexual relations with men outside marriage, and hardly ever used oral contraceptives. Among these groups, the data indicated a shift towards synchrony of 20 to 25 per cent. The Wellers then went back to lesbians, studying thirty couples, and found no sign of synchrony. Combining their results with those of other published papers, they concluded that overall, the literature failed to demonstrate menstrual synchrony in lesbians. In the light of this conclusion, and the modesty of the effect in the Bedouin study, they suggested that 'prolonged and very intensive contact may not be conducive to menstrual synchrony' after all.[34]

The other major reason why scientists have their doubts about synchrony is that it has been an effect without an established cause. In her original paper, Martha McClintock suggested that it might result from the action of pheromones, hormones which are broadcast through the air to act on other individuals, instead of remaining within the body that secretes them. Although pheromones have become a household word, they have done so despite the absence of proof that they actually exist in humans. It was not until 1998 that McClintock published a paper claiming to deliver this proof, by showing that pheromones from one woman (from her armpit, to be precise) could exert an effect on another. Samples taken from women in the early phases of their cycles shortened the cycles of women who inhaled them, while samples from women at the point of ovulation lengthened other women's cycles. These experiments met one of the main objections which the studies outside the laboratories had faced, that a substantial proportion of cycles in a group of women will align with each other by chance.[35]

McClintock's paper was hailed as the inauguration of a new research programme, offering all manner of insights into human behaviour. There were suggestions that feelings from sexual attraction to xenophobia might be triggered by chemicals which people cannot consciously smell.[36] Instead of 'The Naked Ape', welcome to 'Your Life As A Dog'. As far as the origins of culture are concerned, though, her results are useful as evidence that a capacity for synchrony evolved at some stage. The Wellers' reconsidered views about close contact may also be helpful to the case. Pleistocene females did not sleep in dormitories or work in offices. They spent most of their waking hours in the open, and probably slept in spaces which were covered rather than enclosed. In France, for example, the rock shelters show signs of habitation, but although people painted the deep caves, they did not live in them. If menstrual synchrony research now moves away from situations of close contact, it may be more likely to identify mechanisms which could plausibly have operated in ancestral environments. Such mechanisms need not be powerful today. The selection pressures behind them would have slackened once they had helped establish a symbolic order, in which representations of menstrual blood assumed more importance than the real thing.

Menstrual blood signals that a woman will soon be fertile. To a male hominid, it would sort females into two classes; those who were cycling and those who were not. Although the fertility status of some would be obvious, because they were visibly pregnant or had infants attached, others might be infertile because they were in the early stages of pregnancy or the later stages of lactation, or in poor health. Together, the currently infertile would probably constitute a large majority; while any female who menstruated would become a centre of attention. Males would be alert to blood because it signified both life and death.

It would not, however, help them to pursue an ultra-male strategy of maximizing the number of mates and minimizing investment. In order to benefit from the signal, a male who homed in upon a female because she was menstruating would have to guard his position against competitors at least until she reached the point of ovulation. Since he would not know when this had occurred, it would be in his interests to stay rather longer. While she had his attention, his mate would be in a strong position to bargain for signs of commitment, and his responses would help her to decide how good a prospect he really was.

At this point, according to Camilla Power, deception enters the picture. As a means of leveraging male energy, menstruation would be so useful that females would try to fake it. A female could sham menstruation by daubing herself with the menstrual blood of others, which it would be in her sisters' common reproductive interests to provide. In this shared deception is the germ of ritual.

Sham menstruation would serve to safeguard the interests of the majority of females against possible competition from individuals who were in a position to trade on the resource of potentially fruitful sex. In the scenario developed by Power and her colleagues, a female who begins to menstruate is immediately claimed by the others in the group. They assert bonds with her, by painting themselves in menstrual red, and thereby assert influence over her. She could harness more male labour power than the others by trading sex for it after her menses finished, but this would not be in her longer-term interests. For much of her life, she would be in the majority position; infertile at just the time she needed male assistance most, and for the same reason. If she had gone it alone, when her sexual currency was at its most valuable, she would not subsequently be supported by the other females when her ability to trade on this resource became limited. As disloyalty would not pay in the long run, the game of sham menstruation could go on. Menstrual signals would work in concert with synchrony, instead of disrupting it.

Although the sham would have to be convincing, it would not have to be realistic. A male might suspect which females were really menstruating and which were not, but he would not be able to do anything about it. The message of the sham was that the menstrual coalition had established a solidarity which was not worth challenging. It was a deception of a kind unknown among modern non-human primates, in having a collective rather than an individualistic basis. 'As such,' observed Camilla Power and Leslie Aiello, 'it represents a vital step towards sustaining an imaginary construct and sharing that construct with others – that is, dealing with symbols.'[37]

To convince its intended audience, the deception had to be expensive rather than accurate. The more the signal was amplified, the more believable it would be. Females could build it up by making a noise, gesturing, or using substances that would amplify the message of menstrual red. This was the function of the red ochre that is such a striking feature of early modern human sites. In the process of becoming human as we know it, females invented cosmetics.[38]

The process of becoming anatomically modern continued the expansion of the brain, ratcheting up the costs of reproduction that females had to bear. They became less inclined to range widely in foraging parties, encumbered as they were by increasingly dependent young, and they began to remain at home bases in order to conserve energy. Whereas sham menstruation had worked by encouraging males to hang around, in the hope of guarding mates successfully, this now became a disadvantage. Females needed more male labour power. Increased productivity was indicated. The males had to be induced to embark on hunts that were sustained until success was achieved.

Gradually, the sham became detached from hormones, and settled into a rhythm of its own. Perhaps all it took was a little synchrony and a lot of amplification; perhaps they were even able to entrain their rhythms to the phases of the Moon, whose cycle coincides so uncannily in length with those of women. It became a monthly ritual, cued by the Moon, regardless of whether anybody in the coalition was menstruating or not. The female coalitionists had created something whose meaning was not tied to a physical referent. This primal abstraction became a 'morally authoritative intangible', through which right and wrong behaviour could be ordained.[39] It was the first step into the imaginary world, and towards the gods.

Now the ritual served to impose a monthly rhythm under which labour was divided between the sexes. Women's displays, loud and vivid and emphatic, inverted the normal message of sexual assent. To confirm the possibility of mating, an animal needs to verify that the potential mate is of the right sex and species, and that the time is right for fertile sex. The message of the women's ritual was 'wrong sex / wrong species / wrong time'; or, in a word, 'No!'. They were refusing sex, collectively, unless men went out to hunt and returned with provisions. Individual contracts were not to be permitted to breach this solidarity. Although monogamy was favoured as the underlying relationship between the sexes, the ritual urged the men and the women to go their separate ways until the hunt was done and the Moon was full.

In order to deter the hunters from eating some of their kills in the bush, the network of beliefs drew on the power of blood to proscribe eating raw meat. Women, consolidating their home bases, were in a position to assert control over cooking, the process through which meat ceased to be bloody. By creating the symbolic distinction between the raw and the cooked, they exerted influence over men at a distance.

This was the strike that launched culture, and also the birth of taboo. Menstruation, blood, meat and sex were linked into a network of powerful ritual laws, whose traditions endure to this day.

Archaeologists have identified threads of continuity in South African San culture that stretch back 25,000 years. Knight and his colleagues have traced them back even further, to culture's inaugural rituals. They noted the flowering of a recognizable culture that seems to follow the emergence of modern humans, with their sudden appetite for red ochre, beads and other signs that their lives had become more than practical. The new way of life is first seen in Africa, from around 50,000 years ago.

Its founding traditions remain at their strongest, according to Knight, among Africa's last hunter-gatherers. The taboos are still widespread among San groups and the Hadza, hunter-gatherers who live in Tanzania. Not only are San men discouraged from having sex while their wives are menstruating, or from hunting at that time, but abstinence is also prescribed before they go off to hunt big game, or if they are about to resume tracking a wounded animal. Hadza men are also warned not to have sex or hunt during their wives' menstrual periods. The Hadza believe that full moon is the best time for hunting in the dry season, and that women align their time of menstruation to the dark moon. Hadza groups come together during the dry season, and on each night of the dark moon, all fires extinguished, they hold their most important religious ceremonies. One of the themes of *epeme* dances is the resolution of the sexes' conflicting interests, which are elsewhere obtrusive; the festival is considered vital to ensure good health and successful hunting.

The colour of blood is given ritual significance, as when a 'new maiden' of the /Xam (a San group persecuted to extinction in recent times) marked her first menstrual cycle by presenting all the other women in her band with pieces of blood-red haematite ochre, which they used to colour their faces and cloaks. In the Eland Bull Dance, San women surround the new maiden and present their buttocks to her, pantomiming the courtship behaviour of eland cows. The association of 'wrong sex' and 'wrong species' with 'wrong time' is underlined by the Zu/'hoasi (!Kung) identification of the new menstruant as the 'Bull Eland'.

Though the more recent accounts of the 'Human Symbolic Revolution' have concentrated on southern African ethnography, resting as it does upon 100,000 years of ochre, *Blood Relations* describes an imaginary world that spans the real world, installed by humans as they explore one part of the globe after another. Even in the last continent they reach, they continue to paint themselves with red ochre and pigment, so that when explorers from another continent happen upon them thousands of years later, they become known to the newcomers as 'Red Indians'.

In the process of the imaginary world's expansion, its stories evolve and its fauna mutates. Australian Aboriginal myths tell of a Rainbow Snake which creates the world. (One of its names, Uluru, is becoming known to tourists as that now preferred for Ayers Rock.) Knight is scornful of Western scholars' attempts to confine the meaning of this uncontainable symbol into compartments such as 'water', 'phallus', or even the water-python *Liasis fuscus* Peters [sic]. A snake is the most liquid of animals; it flows like water, or menstrual blood. It is the most elastic, swallowing objects whole. In the Rainbow Snake, these qualities assume supernaturally limitless proportions. In Knight's words, it is 'paradoxical to the core', both male and female, occupying the heavens and the deep. It seems to be the great intangible that can enfold all the contradictions of the human world, and as such is the descendant of the first intangible, signifying fertility, morality and ritual power.

6

Fantastical ideas are the kitsch of human origins. One group of e-mail discussion forums specifically prohibits 'Bizarre Theories', along with creationism, racism and rudeness, because these are the things that are fatal to a constructive discussion of how we came to be what we are.[40] Many of Chris Knight's peers seem to have assumed that his is just another wild hobby-horse, and have therefore let it pass them by. Others may be familiar with the ideas but do not know quite what to do with them, as one reviewer observed.[41]

Robert Foley and C. M. Fitzgerald have paid Knight's theory the compliment of treating the adaptive value of synchrony as a testable hypothesis.[42] They conclude, however, that it is an unlikely one. With a touch of Darwinian one-upmanship, they observe that their computer simulation introduced 'a measure of the costs involved' for ancient hominine females who synchronized their cycles. Their model set a probability of infant mortality, causing offspring to die each month. The mortality rate ranged between 50 and 40 per cent in the first five years of life –

in the real world, a rate of 30 per cent has been recorded among the Hadza, and of 30 per cent in ten years among the Zu/'hoasi or !Kung. By dying, the offspring provided opportunities for their mothers to reproduce again before the time set by the group's collective cycle. In the terms of the model, this is 'cheating', by stealing a march on the group.

Foley and Fitzgerald found that females who cheated, by not waiting until the collectively determined time, would end up with more surviving offspring than those who maintained group synchrony. The higher the rate of infant mortality, the greater was the advantage of cheating. Synchronizers only reproduced more than cheaters if the infant mortality rate was below 15 per cent. To achieve a rate that low, females would probably have secured high levels of paternal investment already. Under conditions in which reproductive synchrony could evolve and remain stable, it might not be necessary anyway.

Camilla Power, Catherine Arthur and Leslie Aiello replied that the model Foley and Fitzgerald had tested was not in fact Knight's theory at all, but Paul Turke's model of ovulation synchrony. They mounted a defence all the same, countering that the conditions of the test had been unrealistic. Foley and Fitzgerald had envisaged grand synchrony, with at least seven out of ten females reproducing in concert. Power and her colleagues objected that it was hard to see how female hominids could achieve synchrony at such a level without harming their reproductive interests. If a female in Foley and Fitzgerald's model were to lose a child, and wanted to obey the synchrony rules, she would have to suspend sexual contact for a long period. This would prevent her from harnessing the powerful energy source of 'mating effort' – the effort males are prepared to make in the pursuit of copulation. 'Males who could not gain fertile matings within a band of females for several years might be inclined to search elsewhere rather than persist in mating effort,' observes the reply, and indeed they might.[43]

To produce a model in which synchrony is a stable strategy, Power and her colleagues at University College, London, introduce an element of seasonality. If births are concentrated in a particular season, the costs of synchrony are reduced. A female who loses an offspring has only to wait until the next breeding season, not until all the other females are ready to reproduce again. Nor are mysterious forces needed to impose a seasonal bias on births. Food shortages appear to have this effect, as may periods of exertion that deplete the energy available to women for sustaining pregnancy. Among the Lese, who rely upon the gardens they plant in the Ituri forest of north-eastern Congo, formerly Zaire, conception rates decline in the period following the harvest. But no such effect is seen among the Efe, their pygmy neighbours who live by foraging.[44]

In the seasonal version of the model, female synchronizers only gave birth during a window of three months each year. The infant mortality rates ranged from the 'low' 30 per cent in ten years of the Zu/'hoasi to 54 per cent in ten years, as recorded among chimpanzees. Under these conditions, synchrony was much more viable. The UCL researchers concluded that it could become a stable strategy if synchronizers could secure a reduction of infant mortality of less than 5 per cent; which they suggested could be achieved through the concentrating effect of synchrony on male attention, or by making the most of food resources through giving birth at the optimal time of year. The refinement of seasonality has now been incorporated into the theory. Foley remains unpersuaded, objecting that humans are not seasonal breeders in the sense that the simulations demand.[45]

Knight and his colleagues make it clear that this is how they want to proceed. Their theory is to be treated as science, not as a work of imagination. It should be assessed and modified on the same criteria of truth as it would if it dealt just in biology. Commenting on one of their papers, Robin Dunbar praised them rather faintly with the observation that 'it is often more important to be interesting than to be right'. 'But we are not too interested in wrong hypotheses,' they retorted. 'Had we been wrong – had females not "literally" pursued the strategies we model – we doubt we could have seemed interesting at all.'[46]

They would not have learned their own lessons if they had responded otherwise. When the interests of signaller and receiver do not coincide, the signal must be loud if it is to overcome the receiver's scepticism. According to the theory, the menstrual coalitionists had to amplify their message with noise, dance and red pigment. According to scientific protocol, the equivalent procedure for the theory's proponents is to declare the soundness of their science as forthrightly as possible.

In doing so, they turn its very improbability into a virtue. The fact that the theory's predictions are so specific and exceptional, they argue, makes it easier to test. 'Our model will fall under the weight of positive evidence it cannot allow – examples being pre-hunt rituals prescribing indulgence in marital sex; menstrually potent women cooking meat; rock art traditions focused on the human pair-bond,' they declare. 'We await falsification of the predictions our model actually specifies.'[47]

By asserting this proud positivism over ethnographic and archaeological data, they show how radical their programme is. Anthropology, they insist, is capable of generating testable hypotheses. Cultural evidence can be taken and subjected to the same kind of procedures as biological data. If anthropological material is admitted as sociobiological evidence, in order to improve on 'Darwinism's simplistic treatment of symbolic data', anthropology also will be transformed.

Symbolic data, however, are often at the mercy of interpretation. It may be that a southern African rock painting depicts a menarcheal girl in a ceremonial shelter, or that the wavy band joining two Australian rock figures between the legs represents menstrual flow. Proof is impossible, though, and acceptance depends on the shifting weight of opinion in rock art theory.

The symbolic revolution theory also has to address the question of whether some of its key evidence, the ochre, is symbolic or not. Ian Watts has assembled a strong case for symbolism through his exhaustive study of ochre archaeology in South Africa and elsewhere. Various practical roles for ochre have been suggested, such as insect repellent or antiseptic, but one which has attracted particular attention among archaeologists is that of hide preservative. The metal oxides of ochre may block the action of collagenase, the enzyme which breaks down the stuff that holds skin together.

One metal oxide should be as effective as another, though. If early humans were collecting ochre to stop their cloaks and windbreaks from rotting, they should not have been fussy about the colour. Instead, as Watts shows, the southern African deposits show a preference for red over the other colours available. Judging by studies of hide use in the Kalahari, it is also doubtful whether there would be much point in trying to slow down bacterial decay, because hides would probably succumb to wear and tear long before they rotted. Moreover, if ochre was part of a clothing and covering industry, it should be more prevalent in deposits laid down in a cold climate, but there is no such pattern in its distribution.[48]

Watts's study of ochre illustrates the potential of the ritual-symbolism theory to inform specific archaeological and anthropological issues. The theory's principal and compelling impact, however, is in defining the problem that must be addressed by any theory of how humanity in its recognizable form came to be. This is much more than being merely interesting; but neither does it require the literal truth its authors claim.

Robert Foley has remarked that primatologists are particularly inclined to what he calls 'vacuum' theories of human evolution.[49] Ignoring the context in which evolution takes place, they deny that there is anything to explain, or that any fundamental transformation has occurred. We have no uniquely human traits, they aver, just ones that are present in other primates, writ large. Thus the Last Common Ancestor used tools at the same kind of level as chimpanzees, who poke sticks into termite nests, and hominids gradually worked their skills up through stone-knapping to silicon-etching. Knight and his colleagues are justifiably scathing about theories which envisage human evolution as an accumulation of little improvements, through which hominids found themselves able to do all the things that humans can do. The imperative truth they assert is that any theory of human evolution must show how human phenomena, such as culture and language, can be stable evolutionary strategies. Whether these phenomena developed piecemeal or in accelerated bursts, humans did not just drift into humanity.

As Knight points out, Darwinian theory shows that cheating is likely to result in higher fitness than co-operating – and the greater the rewards of co-operation, the greater the unearned benefits to the freeloader.[50] Any theory of how language, symbolism or culture originated has to show how a system based on co-operative agreement could have developed without being destabilized at any stage by the pursuit of individual interests. This, not the exotic content of the menstrual ritual theory, is the criterion for any rival account.

7

Brains trebled in size over three million years, the curve of expansion rising particularly steeply in the period during which *Homo sapiens* emerged; the phase in which the hand-axe handicap model winds up and the female ritual theory begins. The best working assumption is that they grew mainly in response to the challenges they set each other, and so it follows that the more of each other they had to deal with, the larger their brains would be. There are advantages in large groups, such as improved protection against attack from predators – or among primates, from other groups – and perhaps improved efficiency in hunting or gathering food. If there were benefits to be had from coming together in larger numbers, though, hominines would have had a price to pay in expensive brain tissue.

Robin Dunbar has found a strong correlation in primates between mean group numbers and the size of the neocortex, the most recently evolved part of the brain, relative to the rest of the brain. For the species, the neocortex indicates the cognitive ceiling to the size of the social groups that its members can maintain.

For an individual primate, neocortex size limits the number of other individuals with which it can maintain social relationships.[51]

Its principal means of doing so is to groom the individuals it favours, a sensually pleasurable experience for the recipient (even across the species barrier between monkeys and humans, according to Dunbar). This requires not only effort but time; up to nearly 20 per cent of a group's waking hours. Primates appear not to be able to increase this percentage any further, presumably because the rest of their time is taken up with life's other essentials. Dunbar proposes that language evolved as a kind of virtual grooming, which can be applied to several individuals at once, allowing humans to maintain larger groups. And he argues that the stuff of their conversation was gossip, the verbal equivalent of grooming. Like their descendants, the earliest people were interested above all in each other.

Dunbar points out that grooming sends a message of preference – about who to groom, and of commitment – in the time grooming takes. Camilla Power argues that this exposes a flaw in Dunbar's model. With the power of speech, a hominid can attend to three relationships simultaneously for the price of one, 'but the very fact that you can chatter to three people at once reduces the indication of commitment to each grooming "partner" to a third'.[52]

The other thing wrong with gossip is, of course, its unreliability. Although gossip about third parties may be highly valuable, it offers more opportunities for deception than perhaps any other form of communication. Listening to it could become a costly exercise if every piece of information had to be checked. Conversation about other people would have to be limited to the immediately verifiable – lending indirect support to the suggestion, mooted by the actress Lily Tomlin, that the first sentence ever spoken was 'What a hairy back!'.[53]

Power suggests that it was sham menstruation practices that provided females with the necessary basis of trust, since these locked their participants into mutual arrangements that had to be sustained for a long time if the benefits were to be enjoyed. Menstrual ritual was, in fact, the equivalent of grooming.

Speech and ritual have opposite qualities. Ritual deals in repetition and invariance, whereas speech is a means to create novelty. Words are cheap and soft. They correspond to a style of signalling that John Krebs and Richard Dawkins have described as 'conspiratorial whispering', in which senders and receivers have reached an agreement to lower the cost of their signals.[54] As well as being costly and loud, ritual is anything but a matter of agreement. It is coercive and exploitative, designed to overcome resistance, and has nothing to do with the fair exchange implicit in whispering. 'Ritual, like warfare, cannot afford to assume that there are any rules,' Chris Knight observes.[55]

Ritual also has a direct relationship to warfare. In creating a collective representation, like 'Eland Bull' or 'Rainbow Snake', it creates group identity. Developing this identity by aligning the interests of its members, it directs hostility outwards. In-group solidarity is built at the expense of out-groups. With a host of mutually hostile grouplets, internally united and mutually divided by the colour red, the early cultural landscape would have borne something of a resemblance to the far Left of recent times.

Among the means that ritual can employ to overcome its audience are night, fire, dance, drumming, chants, hallucinogens and sexual display. It may use violence, too, in representations of sacrifice or sacrifice itself, and in initiation rites. Initiatory pain reaches its apogee in the widespread Australian Aboriginal practice of sub-incision, which involves cutting along the length of the penis on its underside, up to the urethra, and then flattening the organ out. A boy thus wrenched into manhood paid a visibly high price to become part of his symbolic community. His commitment assured in this way, he is worthy of trust. He will not need to make his everyday signals costly in order to be believed.

Although initiation rites allow costs to be paid in advance,

religions typically require their adherents to affirm their commitment by frequent and regular ritual practice; a principle better
appreciated by Muslims, obliged to pray five times a day, or
Orthodox Jews, with their complicated and demanding obser-
vances, than by the virtually deconsecrated Anglicans who glance
at their watches if the sermon lasts more than five minutes.
Several religious traditions link language to divinity. In the beginning was the Word, says the Bible; the Indian deity Indra is said to
have created articulate speech; similar themes occur in Norse
mythology, while Plato has Socrates saying that the gods gave
things their proper names. And each dollar bill links divinity to
reliability, affirming that 'In God We Trust'.

Knight compares ritual to a bank and words to banknotes. The
bank's authority gives value to the pieces of paper it issues, which
are worthless in themselves. He also suggests that the Word does
more than authorize words. From ritual, he argues, can come the
whole of language: grammar, cases, tenses and all. The key to
this process is pretend play, which is considered to be integral to
the development of language in young children. A child pretend-
ing holds two meanings in mind at once: the actual events and
objects with which the game is played, and the imaginary things
they represent. An adult holding a religious belief must likewise
understand the world in two ways at the same time, to see a rock
as both a rock and a deity, or to tell water from holy water.

The mimetic displays proposed as the costly precursors of
speech were also a kind of pretend play. With a symbolic register
established by the grand pretence of ritual, it became possible to
reduce the costs of the mimetic displays. Pretend-play routines
could be abbreviated, to the point where they became single
utterances, or words. Some went further, to be truncated and con-
ventionalized into grammatical markers. Now that people could
exchange ideas about things which were not physically present,
either at the time or at all, there was a need to create signs indicating tense and case. At the same time, a new momentum

developed, as people became interested in knowing new things in new ways. Language was elaborated by metaphor, the process of making one thing stand for another. The process will continue as long as people speak, striving against the dulling effects of familiarity to hold their listeners' attention.

Classical Greek drama grew out of ritual, as Knight notes, echoing the original drama in which the creativity of speech was made possible by ritual, its opposite. And if he is right, then at every modern theatrical performance, a shadow play re-enacts the beginnings of language.

8

As well as explaining what we cannot fail to recognize as symbolic behaviour, a theory of symbol origins should have something to say about the enigmatic marks which survive from before the great flowering that came with modern *Homo sapiens*. The idea of a symbolic order based on trust does not imply that all symbolic behaviour was impossible until it was established. By allowing symbolic behaviour to be shared, though, the framework of trust allowed it to be sustained.

It seems likely that the Berekhat Ram 'figurine' at the Golan Heights was deliberately inscribed, more than a quarter of a million years ago, and possible that its maker had perceived a resemblance between the piece of rock and a human form. The act of scoring the pebble to make it look as if it were human could be seen as a moment of pretend play. But without an external system of support, such insights would have been symbolic mayflies, dying the day they were born. Each would be a private initiative, which might enjoy some local success, but would inevitably peter out.

A few distant outliers like Berekhat Ram apart, we are really back to the vexed old question of how human the Neanderthals were. Their brains were as large as modern ones, if not larger; they seem to have buried their dead; some of them seem to have collected objects without practical value, and some showed an appreciation of personal ornament in the modern style. The strongest hint of symbolism is in the burials, since the ornaments were associated with late Neanderthals who probably encountered modern humans. But Neanderthal burials were plain affairs. Without unarguable evidence of ritual behaviour, the parsimonious assumption is that they did not represent religious ceremonies. They may well have been more than a way to dispose of a body, though. There are easier ways to get rid of a cadaver, but making it invisible might help to reduce the distress felt by those with an emotional bond to the deceased.

Mostly, the Neanderthals have left no traces of artefacts other than practical ones, and there are no signs that they developed any forms of symbolism peculiar to themselves. The obvious inference, and one which is currently in scientific favour, is that they were just that bit less cognitively capable than modern humans, who had a competitive edge in their ability to plan and organize their activities.

An evolutionary account of symbolism offers a way to make sense of what we know about Neanderthal capacities, without necessarily inferring that their disadvantage relative to modern humans was innate. They were able to meet the costs of large brains with a suite of adaptations that did not include symbolism. Modern humans developed an adaptive package which proved more effective, but it may have been more like the later behavioural shift towards farming, which increased population densities without the use of a mutation. As the menstrual synchrony model suggests, the path to symbolism may have been a tortuous one. Perhaps Neanderthal females just did not hit upon it, or make the right moves when they did; whereas modern females struck lucky, and capitalized on their advantage. But as long as Neanderthals were on their own, their strategies were viable. There was nothing inevitable about becoming human in the modern sense.

FOUR

Benefits

1

A bird is constrained by its wings, which it cannot use to grasp, pull, throw, slash, pick, poke, stroke, squeeze, dig, climb or walk. On the other hand, so to speak and with a few celebrated exceptions, it can use them to fly. Being the stuff of human dreams, this ability is so enviable that it seems perverse to think of wings as constraints. If there were a species of bird in which the pair of limbs nearer the head had evolved into avian arms, adapting to life on some island where flying was not worth the candle, we would be struck rather than impressed by the results of natural selection in this case. Even if the birds could perform a wider range of actions than their ancestors, or perform actions more effectively than they could have done using feet and bills, they would seem to have been limited rather than liberated.

A motorist constrains the motion of a car by the use of the steering wheel. Not only is this good for the survival of both parties, but it is the key to the goal of reaching the intended destination. Another word for constraint is structure. Although the constraints described by evolutionary theory have an obvious appeal for those who want to declare limits to human possibility, they are better understood as what makes any of our actions possible.

Like many of their lay sympathizers and popularizers, evolutionary psychologists generally seem readier to talk about limits than possibilities. They sometimes use the argument that knowledge is better than ignorance, that we should have a scientific understanding of the downside of human nature. This fits with a posture of disinterest, in which scientists say that they are just reporting the news, and it is up to others to write the editorials. Richard Wrangham and Dale Peterson ended their book *Demonic Males* with an appeal to human wisdom, passing the buck to the abstract. Subtitled *Apes and the Origins of Human Violence*, the book is based on Wrangham's experience as a primatologist, much of which was gained observing the chimpanzees at Gombe. Wrangham concluded that humans are cursed with a 'demonic male temperament' and the '5-million-year stain of our ape past'.[1] It is easy to believe that this is how one is likely to feel after a quarter century spent studying chimpanzees. The token reference to wisdom vanishes in the overwhelming gloom.

Wrangham might have felt somewhat more cheerful if he had been watching bonobos, *Pan paniscus*, rather than *Pan troglodytes*. In recent times they have brought much comfort to scientists with their female solidarity, their polymorphous sexual promiscuity, and their tamed males. Before them, Western scholarship had Samoans, portrayed as freely loving, sexually equal and peaceable by the anthropologist Margaret Mead. They were flower children, whereas bonobos reflect the values of a subsequent counter-culture, with their emphasis on homosexuality and female power. They have also never recovered from the assault on Mead's scholarship mounted by Derek Freeman in 1973. Samoa was not an earthly paradise, it transpired, but a familiar kind of place where violence was common, virginity was prized, and adultery punished by death. Now bonobos are the new Samoans, the females among them 'active sexual participants, creatures making decisions about their own reproductive futures', in the words of the anthropologist Meredith Small.[2]

Richard Wrangham does know about bonobos, of course, and he devotes a chapter of the book to their story, which he sees as a tale of 'demonism vanquished'. He attributes the bonobos' roots to herbs which grow on the floor of the forest where they live. Similar herbs sprout on the forest floor trodden by common chimpanzees as well, but they are eaten by gorillas. The bonobos do not have gorillas to compete with, probably through some chance of climatic change and geography, so they can make use of the herbs as a food source as they move through the forest. As their resources are more evenly distributed than those of common chimpanzees, they can sustain larger group sizes. Females are not forced to disperse so widely, and so can form stronger relationships with each other.

This is the significance of sexual contact between bonobo females, known to local people as *hoka-hoka* and to primatologists as genital-genital rubbing. Junior females form bonds not with kin, but through sexual activity with older females. Bonobos also order their heterosexual relations differently from their *troglodytes* cousins. Females seem to have acquired some ability to conceal their ovulation, since males do not show much preference for mating with them at this point in the oestrous cycle. The level of competition and conflict among males is reduced, because they do not fight over opportunities to mate with particular females at particular times.

Easy as it may be to assume that the dominance of violent males is the natural condition, it is not certain that *paniscus* evolved from a creature more like *troglodytes*. But it is quite likely, and if so, the transformation was reformist rather than revolutionary. Females leave their family group, while males remain, which is how things are among common chimpanzees. The difference lies in the bonds that females form; beyond kin, with unrelated females, and among kin, between mothers and sons. Male bonobos will still be males, and they still form associations with their male kin, but they have spats rather than bloody contests. They do not conduct wars, either: female bonding appears able to bring groups together as well as individuals.

Regrettably, this idyll depends on the forest and its herbs. Wrangham and Peterson doubt that food sources were spread evenly enough through the savannah to support female bonds of such strength in the hominid lineage. Among palaeoanthropologists, 'savannah' has become shorthand for a range of environments containing different proportions of grassland and woodland. The larger the woods, the better the chances would be for hominid females to find food in large patches, lowering the pressure on them to disperse. The South African palaeoanthropologist Lee Berger argues that the long arms and short legs of *Australopithecus africanus* were an adaptation to the trees of southern African forests, which may have endured longer than the forests around the northern reaches of the Great Rift Valley. He suspects that the first hominines may have evolved from *africanus* rather than the northern form, *afarensis*. His idea shows human evolution in a more arboreal light than the conventional savannah view, and suggests that the pressure on early female hominids to disperse may have been less intense than previous stories imply.[3]

Whatever the selective balance between woodlands and grasslands, bonobos demonstrate that there is one way to amend a social system of the chimpanzee type, which suggests that there may be others. So does the remarkable increase in size seen with the emergence of distinctively hominine forms, designated *Homo ergaster* or *erectus*. Both sexes grew larger by the same amount, reducing the difference in size between them. Henry M. McHenry suggests that they may have grown bigger because they became able to eat a wider range of foods. If so, an environment with resources in small patches might be transformed into one whose resources were distributed evenly or in large patches, making it easier for females to spend time and develop bonds with each other. McHenry also speculates that competition between males may have decreased at this stage.[4]

Robert Foley notes that *ergaster* occupied dry and diverse areas. He and Phyllis Lee doubt that a shift towards female social bonds took place at this stage. They do, however, allow that conditions may have become more favourable for such a transition later on. Their account of the *erectus* dietary transition presents it more as a shift away from plants towards meat, rather than as a general broadening of the diet, and they emphasize how patchily meat tends to be distributed. They consider that the *erectus* way of life would have favoured conservation of male kin alliances, and competition over territory. Later on, though, with abundant food and rapid expansion of population and territory, Foley and Lee believe that it might at last have become easier to establish bonds between mothers and daughters.[5]

Changes in social relations would surely have taken place as groups and brains grew larger during the Acheulean era, and they must have impinged in various ways on male alliances. Yet even if we accept that traits such as men's aggression have roots deeper than culture, it does not necessarily mean that we just have to take our sociobiological medicine with a spoonful of higher civilization. If we also accept that we are simultaneously biological and cultural, then scientific tools should prove useful in constructing models of the Man Question and other problems.

A case in point, intimately bound up with masculinity, is the idea that humans are innately predisposed to ethnocentrism. This is difficult to support using classical calculations of inclusive fitness. Like the force of gravity, the power of shared genes falls away rapidly with distance. Although it may be immensely

powerful at close range, between parents, children and siblings, it
may be negligible among cousins removed as many times as any
two members of an ethnic group are likely to be. J. B. S. Haldane
said as much in his famous quip, that he would lay down his life
for three brothers or nine cousins. The calculations become perfectly absurd when inclusive fitness is distributed across x million
cousins, y times removed.

The sociobiologist Pierre van den Berghe has proposed that
even as they decline, shared genetic interests remain strong
enough to underpin ethnicity, which forms 'the outer layers of an
onion of nepotism'.[6] This is sociobiology of the old school, always
presuming that humans are devoted to the pursuit of maximum
fitness. A more persuasive argument derives from the historical
approach, which considers humans as the latest in a lineage of
shifting adaptations within the space of possible primate forms.

Primate social space is finite, as Foley and Lee have observed,
and so are the paths within that space. Species have to climb
towards fitness peaks, and cannot go through valleys of reduced
fitness. Once a primate lineage has entered a particular set of
social arrangements, it may have few ways out. Since it seems
likely that the Last Common Ancestor had a social organization
like that of chimpanzees, the hominid die may have been cast:
mutually hostile male groups, fighting over turf. If related males
were in the habit of forming coalitions when the lineage began,
that tendency is likely to have persisted. In that case, the resemblance between bands of male chimpanzees and bands of men
may not be coincidental. Each depends on the sorting of others
into us and them; each is fundamental to the social system of the
species as a whole.

So there is good evolutionary reason to suppose that a tendency for males to form mutually antagonistic coalitions has been
conserved in our lineage. Males are indeed observed to form
mutually antagonistic coalitions at any opportunity. They rarely
need to be told twice to form themselves into teams. The case is

not open and shut, as some evolutionary popularizers would claim, but there is good reason to suspect that we may not have grown out of these particular ancestral dispositions.

This is not a conclusion. It is simply the point at which arguments on these lines are usually abandoned. An ape heritage is a stain that just won't wash out, it is implied, or a price that has to be paid. Matt Ridley, for example, sees the destructive aspects of human 'groupishness' as the other side of the coin of trade. Intimately bound up with 'the origins of virtue', in the words of the title of his book, trade is next to godliness. Whereas a pessimist like Robert Ardrey would finish his expeditions into human origins at a place close to Conrad's Kurtz – 'The horror! The horror!' – Ridley's breezy optimism discovers such virtue in trade as to outshine the darkness in the human heart.[7] End of story, either way.

If left at that, human groupishness retains its primordial aura. Primordial forces are not merely ancient, they are taken to be so fundamental as to be beyond analysis. We are left with an understanding little better than that of the Romantics, who believed that the spirit of a people was the obscure and ineffable given from which its history proceeded. The trouble with most sociobiological treatments of such themes is not that they concentrate on biology too much, but that they do not concentrate on it enough. They assert its influence upon human affairs, but are not interested in going into details. Those are what puts biology in its place.

Male chimpanzee coalitions are like male human ones in being ubiquitous, mutually antagonistic, and often extremely aggressive. They are based on kinship and tend to be unstable, in which respects they resemble some men's coalitions but not others. These differences are fundamental. Chimpanzees are not bound together by symbols. All they have is kinship, not trust. They form alliances, but they do not hold allegiances. If a male chimpanzee sees another band approaching, he must be able to think the equivalent of 'here come enemies'. He must be able to tell ally from foe. But he is unable to think something like 'here comes the Green Banana Troop', or 'I belong to the Tall Tree Gang'. The two *Pan* species make use of leaves and sticks in various ways; as sponges, 'fishing rods' to poke into termite nests, and possibly even as trail markers. But they never adorn themselves with leaves, let alone raise them as flags. If bonobos really do point the way through the forest with sticks and trampled vegetation, they are capable of using objects as signs at the simplest level of reference, where the sign is in physical contact with the object to which it refers. Their trails will never lead to shrines, though, or be haunted by spirits. Without the ability to create abstract symbols, chimpanzees are unable to reify their coalitions as entities with existences independent of themselves. That uniquely human capability is what induces a man to lay down his life not for the requisite number of close relatives, but for his country.

Matt Ridley relates how, in ancient Rome and Constantinople, the assignment of different colours to competitors in the chariot races caused factions to arise in support of each colour. In the eastern city, the 'acid of sporting factions mixed with the alkali of religion and politics'; instead of neutralizing each other – Ridley is a biologist by training, not a chemist – 'they exploded into internecine fury'. In 532, the Greens and Blues united and launched a revolt, proclaiming a new emperor. The Nika insurrection ended with the slaughter of 30,000 rebels in the Hippodrome. 'Nika', meaning 'Win', was what the charioteers' supporters would chant in the stadium.

This episode, Ridley says, 'illustrates that the power of xenophobic group loyalty in the human species is every bit as potent as it is in chimpanzees'. He sells human xenophobia short. It is surely much more potent than the chimpanzee variety. Chimpanzee coalitions tend to be transient and unstable. They rarely attain large numbers for long. It is conceivable that chimps have only a rudimentary group sense. When the number of individuals is small, a coalition could be maintained by each individual's knowledge of relationships between himself and other individual members. The chimpanzee thinks the equivalent of 'that one is with me; that one is with me too', and so on. If all his fellows think the same way, a group will emerge from the sum of their egocentric perspectives. As political primates, they will be keenly aware of certain relationships that do not include themselves: they will be on the lookout for friendships, rivalries and power struggles. Yet none need have a view of the big picture, a sense of the group as a whole.

Whether or not chimpanzee minds are capable of integrating enough relationships to form a true group sense, hominines must surely have been able to do so long before they had abstract symbols. As groups became larger and social dynamics more complex, such an ability would have moved well up the adaptive agenda. Even with cognitive systems specialized for the purpose, though, the faculty would be limited by an individual's ability to recognize other individuals. Individuals would be more likely to perceive a hominine gathering as 'a group of beings like me', rather than 'many beings like me'. After the gathering had dispersed, though, they would have no means of maintaining the perception, except in so far as they were able to make a pattern out of relationships between individuals personally known to them.

With symbols, human group sense was liberated and multiplied indefinitely. A chariot-racing fan in Constantinople or a Communard in red Paris could recognize one among thousands of strangers as an ally by the sign of colour, an arbitrary signifier denoting an abstract entity. In both cities, as in innumerable other places, the consequences proved cataclysmic. The innate components of xenophobia may be stronger, weaker, or about the same in humans as they are in chimpanzees. But human coalitions are more numerous, durable and powerful than those of chimpanzees because they exist through symbols, which have a life of their own.

So far from denying biology, or the ape heritage, this argument recognizes that we remain primates and may have inherited a disposition to the coalitionary behaviour observed in the male of the species. By following the primate logic through, in comparisons with chimpanzees and by constructing the evolutionary story that accords best with our current knowledge, the argument identifies symbolism as what turns family groupishness into superconductive ethnic passion. It suggests that instead of bemoaning the supposed ape within, we should work out ways of living by the symbol which do not lead to dying by the symbol.

For one thing, we can take an optimistic view of the national question. Contemporary musings about the inborn roots of xenophobia make frequent reference to nationalism and tribalism, the latter loosely denoting a primitive clannishness. But according to kin selection theory, the only group that definitely enjoys innate privilege is the family. Although a clan or a tribe is based upon kinship, the much weaker bonds of inclusive fitness between distantly related members must be shored up by social structures. Larger groups depend still more heavily upon their customs, laws, culture and symbols. Contrary to the fundamental tenets of Romanticism, there is nothing primordial about the ethnic group. A nation may command more intense loyalty if it corresponds closely to an ethnic group, but the latter's gravitational attraction arises from familiarity and history, not shared genes. And since humans are able to assign items to more than one category, they are capable of assigning themselves to more than one group or institution. There may be many political or cultural obstacles to the creation of a supranational European identity, for example, but biology is not standing in the way. Nations are symbolic, not natural.

Sometimes sociobiological arguments cite the prevalence of kinship metaphors in national or ethnic rhetoric, implying that if

a man calls another his brother, he will be encouraged by the kinship instinct to treat that man like a brother. Of course he knows
that the other man is not really his brother, but the power of the
word is so great that it entrains the kinship machine anyway. Part
of him sees his fellow as a brother, in the same way that people
still see the Moon as bigger when it is near the horizon, even
once they understand that it remains the same size.

If there is such a thing as a kinship module, then it should
indeed respond to words as it does to any other indicators of relationship. But that would not necessarily give it any significant
influence upon thought or behaviour, any more than the Moon
illusion gives people hope that the satellite is coming within
reach. Indeed, the extent to which kinship is invoked may often
bespeak the metaphor's weakness. African-American men speak
of 'brothers' as an expression of an ideal solidarity which they are
far from achieving. Russians speak of their 'brother Serbs', but are
not inclined to translate sentiment into commitment. When the
metaphor is imposed, it may be entirely devoid of influence. 'I
can't go and visit my brother,' a Pole once remarked to me. He
wasn't complaining, exactly. This was during the Soviet era (and
he was my uncle, since we're on the subject). He was making the
ironic observation that despite the rhetoric of socialist brotherhood, the authorities went to enormous lengths to place obstacles
in the way of ordinary Poles who wanted to enter the Soviet
Union. Not many of them did, because the relationship between
Poles and Russians is more like that between cats and dogs than
kin. Although the shared Slav identity is necessary for brotherly
feeling, it is conspicuously insufficient. Slav brotherhood is a sentiment based more on shared religious affiliation, such as the
Orthodoxy of Russians and Serbs.

According to standard procedure, the argument would at this
stage be deemed to have passed into the social domain, like a
property changing hands. But it has not simply eliminated biology
from its equations. Instead of ordaining that this problem belongs

to the humanities and not to science, it seeks to sketch a map
showing where biological features may stand in relation to social
ones. Evolutionary thinking can be used for ends more positive
than specifying limits to human possibilities. Darwinism can also
be used to place limits on pessimistic claims about human nature.

2

At the second Evolution of Language conference, held in London in 1998, Geoffrey Miller gave a paper in which he argued that sexual selection is the reason there are so many words in the dictionary. Literate adults typically know about 60,000 of them, but just 100 words account for 60 per cent of conversation, and 4,000 words comprise 98 per cent of speech. Miller argued that vocabulary size, like the complexity of bird song, is a reliable indicator of brain size and efficiency. A large vocabulary is also, he proposed, an aid to aesthetic display in courtship. With an amply stocked lexical bag to draw upon, a suitor can entertain and impress potential mates 'by activating the greatest diversity of concepts in their minds'.[8]

The reaction was sceptical, and interestingly so. Unlike some of the more specialized presentations heard at the conference, this was a paper to which everybody could come up with an objection. Some listeners questioned Miller's heavy reliance upon the concept of general intelligence, as measured by IQ tests. This provoked Miller into a trenchant defence of the IQ concept, and the genetic theories of individual difference with which it is associated. He is one evolutionary psychologist who publicly embraces the dismal science of behaviour genetics, instead of seeking to distance his discipline from it.

For Miller, sexual selection is a first rather than a last explanatory resort. When he spots a pattern in human behaviour that resembles patterns attributed to sexual selection in other species, he gives the selective explanation priority. He is building up his case that the human brain was expanded by the pressure of sexual selection, so as to act as a courtship device.

Another phenomenon he lays at courtship's door is cultural production. Surveying the demographics of artists, musicians and writers, he has concluded that men have produced ten times as much culture as women, and that they are at their most creative during young adulthood, rather than in maturity. This conforms to the lines of sexual selection theory, since women would be expected to invest less in display, having less to gain in reproductive success by attracting more mates. Admittedly, if an anthropologist from another planet asked you to explain rock-'n'roll in three words, 'male sexual display' would sum it up nicely, but this is just the kind of thing the geneticist Steve Jones has in mind when he says that science can answer all questions about matters such as sex and opera except the interesting ones.[9]

To dismiss Miller's ideas on these grounds would miss the point, however. Unlike the handaxe, vocabulary size is very easy to explain without invoking sexual selection. A large and complex society, composed of diverse groups engaged in diverse activities, will accumulate a large lexicon. Individual competence will be enhanced by participation in different activities, and by a communicative culture in which words readily migrate from one domain to another. But Miller's project entails challenging conventional explanations of human behaviour, even when they appear to be satisfactory. 'We must face the possibility that most current theories of human behavior and culture are inadequate,' he has written, 'because they may have vastly under-estimated the role of sexual competition, courtship, and mate choice in human affairs.'[10]

His polemical approach provides a good illustration of why Chris Knight's work is so important. Modern Darwinism began with the disavowal of group selection in evolutionary biology; Miller's radical articulation of evolutionary psychology and behaviour genetics hints at the replacement of groups by selection in the human sciences. It hints that what Émile Durkheim called 'social facts' – entities external to the individual, such as law – are peripheral to the understanding of human behaviour. As far as this reading of evolutionary theory is concerned, Margaret Thatcher hit the nail on the head: there is no such thing as society, only individual men and women and their families.

If Miller had been speaking for the motion at the Great Sociobiology Debate, that Darwinism can explain the origins of culture, it would have been a very different story to the one Knight told – and perhaps a different vote at the end. Many human scientists might be prepared to come to terms with Darwin for individuals, as long as they can keep Durkheim for groups. Few, or none, would accept Darwin all the way. Knight's theories form a prototype for an analytical programme in which Darwin and Durkheim could enjoy a mutually supportive relationship. Its ambitions entail the establishment of an evolutionary basis for social facts.

Bold as it is, Knight's programme is based entirely upon accepted components of mainstream modern Darwinism. They are not what the public commonly associates with Darwinism, though, and they are not what usually features in public controversies about evolution. Instead of asserting the existence of genes for this or that, Knight and his colleagues have avoided the easy option of claiming that suddenly a gene mutated into place and culture was born. Evolutionists who invent genes when faced with a shortage of ideas are like governments which print money when they encounter a shortage of cash. They are apt to cause theoretical inflation, and undermine the credibility of the currency.

Miller was in a more statesmanlike mood when he took part in a discussion at the launch of the Darwin@LSE's *Darwinism Today* series. Responding to Kingsley Browne's conservative reading of evolutionary theory, which argued that sexual inequality in the workplace was the result not of discrimination but of different psychological adaptations, Miller suggested that if you didn't like Browne's evolutionary account, you should write your own.[11] Implicit in this excellent advice are two very important precepts: Work within the paradigm, and realize the freedom you have within it.

Presented cold with an exhortation to embrace Darwinism, many people will suspect that they are being urged to adopt a faith. Evolutionary psychology's self-proclaimed quest to define a universal human nature may be seen as a proposal for a belief system, which provides answers that are not only necessary but sufficient to understand the human condition. This book has been in large measure an attempt to present a different view of evolutionary perspectives on human evolution, as a set of practical tools. For this reason, I have tended to refer to evolutionary theory, in the lower case, rather than Darwinism with a capital D. With Marx and Freud off their pedestals, Darwin may be the last man standing among the great intellectual system-builders of the nineteenth century, but the last thing we need now is another patriarchal'ism'.

One important difference between a religion and a secular body of thought is that one is not obliged to accept the entire package. The proliferation of modern Darwinism can make it difficult to see the paradigm for the theories. Many of these will be what the evosceptics claim: speculative, weak, ideologically inflected, or all three. Others will be better constructed, but also wrong. Some may be speculative and ideologically inflected, but right. What matters is that the core theory proves robust, and it is significant in this respect that most of the sceptical fire has been directed at the easiest targets.

In particular, critics have objected that evolutionary psychology has failed to take proper account of culture, and mistakes current social relations for human universals. They have said that the discipline is undisciplined in its methods, though they do not reject the principles on which it is based. Although the arch-critic of evolutionary psychology, Stephen Jay Gould, writes prolifically on a wide range of evolutionary topics, concepts such as inclusive fitness, reciprocal altruism and sexual selection are not especially prominent among them. Gould's judgement of the core of evolutionary psychological theory is therefore less clear than his disdain for what he considers to be merely 'cocktail party' speculation. He has, however, agreed that the 'recognition that differing Darwinian requirements for males and females imply distinct adaptive behaviors' is the evolutionary psychologists' 'most promising theory'.[12]

Gould's complaint is that the theory has been compromised by 'turning a useful principle into a central dogma with asserted powers for nearly universal explanation'. A dogma is a doctrine which may not be criticized or modified. Scientists are at liberty to criticize the principle of different reproductive interests between the sexes, but so far it has remained secure and has proved productive.

One way of affirming its power without taking it as an article of faith might be to draw an analogy with the physical principles that govern what happens when an electric current meets a magnetic field. Applying these principles, engineers can build electric motors, in all different sizes and for a host of diverse applications. Evolutionary hypotheses and methods are like these motors: they are tools, not sacraments. They will give poor results if used carelessly, ineptly or anti-socially; but in that respect they are no different from any other kind of tool. So here

are some guidelines that should help get the best out of an evolutionary tool-kit.

Evolutionary theory cannot say how strong our instincts are.

Evolutionary hypotheses depend upon numbers and closely specified models to raise themselves above the level of the just-so
story. But when they are introduced into the public sphere of
debate, about social relations, sex differences or morality, they
lose touch with numbers.

It is possible to suggest what evolutionary theory might predict
about an aspect of human psychology, but it is not possible to
specify the strength of the effect. Evolutionary insights are
insights about aspects of the mind presumed to have begun evolving long before the dawn of the symbolic world, in which
humankind has spent the last 40,000 or 50,000 years. In its present
form, evolutionary psychology is unable to give measures of the
strength of the forces it proposes, so that they can be weighed
against those of culture and symbolism.

This strips evolutionary perspectives of their privilege. Though
they base their credentials on science, they cannot use hard data
to claim priority over sociological, religious or any other points of
view. Instead, they have to rely heavily on informal perceptions of
human nature. Where evolutionary theory matches patterns of
behaviour that are widespread and consistent, the theory gains
credibility. Men's hankering after promiscuity tends to be greater
than that of women, who tend to be more careful than men in
their choice of sexual partners. Many men are perfectly happy to
be monogamous, and many women are not; but this is consistent
with evolutionary thinking, which expects to encounter variety.

Anybody wishing to derive norms for sexual behaviour from
evolutionary theory might be disappointed to learn that it can also
embrace sexual pursuits without obvious reproductive potential.
One possibility is that, contrary to intuition, practices such as masturbation may enhance reproductive fitness after all. Bolstered

by genetic surveys from several locations, indicating that one child in five was not fathered by its supposed father, Robin Baker and Mark Bellis have developed a theory about the adaptive advantages of extra-pair copulation. Opportunistic intercourse of this kind, they argue, may be a good way of having the best of both worlds, by obtaining good genes without strings attached, while retaining male investment from a long-term relationship. By drawing acidic fluid from the vagina into the cervix, the muscular contractions of orgasm can kill or impair any sperm already present. Baker and Bellis argue that through masturbation, women can exercise a degree of influence over the outcome of the contest between sperm from different sexual partners.[13]

The effect of masturbation would depend on who the next copulatory partner was, and maintaining the proper schedule might be tricky, to say the least. But although the Baker and Bellis model is convoluted, their vision as a whole has a ring of truth about it. Sexual competition probably is as involved and devious a process at the level of bodily fluids as it is at all the levels higher up. Since it cuts to the quick of reproductive success, it is likely to have generated a number of adaptive behavioural patterns. There is no need to identify all sexual behaviours as evolved competition tactics, though. Some might be for entertainment purposes only. Being curious and inventive creatures, it would not have taken long for hominids to discover that masturbation was pleasurable, as other primates have.

There is a third possibility: that humans, in their inventiveness, have exapted sex for wider purposes than reproduction. The contrast between the social lives of ordinary chimpanzees and bonobos suggests that the latter smooth their social relations with sexual activity, much of which cannot lead directly to reproduction because it is between individuals of the same sex. Bonobos show that primates are capable of evolving dispositions to forms of sex that are adaptive without being reproductive. With much higher intelligence, and much more complex societies, the potential for hominines to adopt similar strategies would be greatly increased, as would the pressure on them to do so. Thus, human sexual variety and pluralism might have an adaptive basis, as a means of consolidating bonds or reducing tensions, between members of the same or of different sexes.

An evolved mind is not a predetermined mind.

The question of evolution and human freedom is the question of how Steven Pinker can tell his genes to go and jump in the lake. That, he declares in *How The Mind Works*, is what they can do if they don't like the fact that he has failed to replicate them. He is happy to be childless, which makes him 'a horrible mistake' by Darwinian standards. He embraces a view of the mind as an accumulation of adaptations, all of which evolved because they tended to improve the reproductive fortunes of their possessors. Yet, after several million years of hominid evolution, one of the finest minds in the species chooses to contemplate reproductive success instead of pursuing it.

Sceptics may sense something wrong here, whether in the adaptations, in the theory of an evolved psychology, or in Professor Pinker himself.[14] There is less to his boast than meets the eye, though. Reproductive interests act principally through their agent, sexual desire, which is spectacularly effective even in the absence of a desire to reproduce. Without indulging in impertinent speculation about Steven Pinker's private life, we may note that the modern environment contains artificial contraceptives which make it quite painless to tell one's genes to jump in the lake.

In any case, an individual behaving maladaptively does not subvert evolutionary theory. Organisms have always behaved maladaptively. As a result they are eaten, crushed, drowned, poisoned or suffer other loss of fitness. They leave fewer copies of their genes behind them, and that is that. Humans have more sophisticated ways of not taking care of their reproductive interests, including many forms of behaviour which are very enjoyable. It is hardly surprising that some of them have chosen to explore the gap that has opened up between self-interest and reproductive interest. Meanwhile the exponential increase in the human population suggests that for the most part, the adaptations are working only too well.

There are also ways of confounding sexual psychology which are not sensually agreeable, and may impose great burdens on those who try to practise them. Many religious precepts confront reproductive interests rather than finding ways around them, and so raise more substantial questions about the power of evolved psychology than do artificial contraceptives. All variations of Christianity, apart from the odd orgiastic cult, prescribe monogamy. In evolutionary terms, they insist that only one of several adaptive strategies is legitimate. Although often honoured in the breach rather than the observance, strict religious codes of sexual morality have undoubtedly enjoyed a great deal of success in editing the evolved codes of sexual practice.

The result can be seen as a symbiotic association between the flesh and the word. At the heart of the relationship is a shared commitment to a reproductive strategy which can be extremely successful, promoting as it does high levels of parental care. The word intensifies the commitment of the flesh to this strategy, and achieves its own replication within it. Families with high levels of commitment from two parents are ideal environments for imparting a belief system to the children. The reproductive successes of the people and the ideas are mutually reinforcing.

This is an example of a view of life which has gained considerable currency in recent years, thanks in considerable part to Richard Dawkins's coining of the term 'meme'. A meme is the cultural equivalent of the gene, a replicating unit which thrives according to how well it fits into its environment. Successful memes include folk tales, jokes and ideas whose time has come. Memes imply that ideas really do have a life of their own. Genes are information-coded in one particular way, but selection can operate on information regardless of the form it takes. If units replicate and adapt, they are exhibiting the fundamental characteristics of life. Electronic replicators have spread beyond the laboratories and are now colonizing Internet servers and personal computers, whose fauna of 'artificial life' forms ranges from mathematical stick figures to virtual pets with saucer eyes and a choice of hairstyles. Frivolous as many of them are, they are serving a significant cultural role, in making the idea of non-biological life familiar.

Memes are not yet part of the system. Even Daniel Dennett, convinced as he is about the unique explanatory power of evolution, is cautious about whether a science of memetics is possible.[15] If such a discipline does emerge, it could transform our understanding of how selective processes operate within our species. We could then ask answerable questions about whether ideas, such as religious beliefs, really can replicate at the expense of their human vehicles. For the time being, though, the approach taken in this book may be more appropriate to the state of our knowledge: start at the beginning, and try to fill in the blank spaces on the way.

Since brain tissue is metabolically expensive to create and maintain, it is not plausible that the brain simply got bigger as a side-effect of some other evolutionary process, creating extra capacity by accident. It is possible that extra brain capacity was useful for a variety of reasons, and so tended to take a general-purpose form. The alternative is the modular view, which is more inclined to see each new brain investment as a response to a specific selective pressure. Cognitive capacities would therefore grow by the addition of new specialized systems. Taken to its logical conclusion, this implies a mind made up of large numbers of instincts. Those influenced by the tradition of the *tabula rasa* may be inclined to equate blank space with freedom, and instincts with constraints.

Freedom of action is limited without flexibility of thought, so the issue of freedom rests upon the question of how cognitive flexibility is best achieved. Leda Cosmides and John Tooby cite the psychologist William James in support of modularity. 'It was (and is) common to think that other animals are ruled by "instinct" whereas humans lost their instincts and are ruled by "reason", and that this is why we are so much more flexibly intelligent than other animals,' they write. 'William James took the opposite view. He argued that human behavior is more flexibly intelligent than that of other animals because we have *more* instincts than they do, not fewer.'

In common usage, instinctive behaviour is not only automatic, but exclusively so. Instinct is something that takes over in emergencies, or tells us how we feel without explaining why. What Cosmides and Tooby are describing is more akin to a battery of routines which may engage automatically, but may also act in the service of conscious reflection. Cosmides has argued that people find logical problems easier to solve when these take the form of questions about breaches of social contracts. She presented her experimental subjects with puzzles that had the same logical structure, but different façades. They were based on a format known as the Wason test, in which the subject is presented with four cards. His or her challenge is to turn over only as many cards as are necessary to verify the relationship between 'if' and 'then'.

Relationships such as these are the coinage of social life, from the point at which children are told that if they have finished eating, then they may leave the table, or if they are good, then Father Christmas will bring them presents. Children rapidly prove themselves highly attuned to breaches of conditional rules, on the part of adults or other children, if not themselves. In Cosmides's experiments, adults showed themselves adept at detecting situations in which individuals enjoy a benefit without paying the cost required by the rule governing the contract.

Their aptitude extended even to imaginary social contracts far removed from their own experiences. Her subjects proved better at evaluating a rule such as 'if a man eats cassava root, then he must have a tattoo on his face', than at solving puzzles about familiar matters which did not involve a social contract, such as whether people always use the subway to go into Boston. Three-quarters got the right answer when the puzzle revolved around cassava and tattoos, but less than a quarter figured out the urban transit question. Scores under 25 per cent are in fact typical in Wason tests, leading psychologists to believe that the human cognitive apparatus is not especially impressive in dealing with 'if' and 'then'. Cosmides concludes that the massive performance boost seen in the social contract tests is achieved through the engagement of specialized mental machinery, adapted to detect cheating.[16]

The automated element is the activation of an instinct – algorithm is a more modern term, with more appropriate connotations – in response to signals indicating that the domain is social and the question concerns fairness. It is as if the mind is prompted to reach for a pocket calculator when its attention is directed to the possibility of cheating in a social contract, but not otherwise. The instinct is a tool which the conscious mind can use to help it make decisions, not an elemental force that lays down the adaptive law.

As for the conscious processes themselves, there is nothing in them that defies natural selection. They may arise from some sort of 'metarepresentational module', to use Dan Sperber's term, which has evolved to integrate outputs from the other modules. Or they may be part of a general cognitive engine which pre-dates the modules that feed into it. Wherever they spring from, they are consistent with the view that the modern human mind has evolved a winning combination of specialized and general systems. Notwithstanding Jerry Fodor's opinion that most cognitive scientists still believe in the blank slate, 'so deeply, indeed, that many hardly notice that they do', it is widely accepted that perception and grammar are handled by dedicated circuitry.[17]

Experimental studies in recent years have suggested that infants are born with a body of basic knowledge about the world. If they are shown illusions that appear to violate physical laws, such as an object seeming to pass through a gap smaller than itself, they will pay more attention than they do to unremarkable sights. They also appear to recognize a difference between animate and inanimate objects, understanding that animate ones are likely to behave in different ways.

The debate is over how high the specialized systems go. Although there has been a decisive shift towards the view that the mind contains a great deal of innate knowledge, support is more limited for claims that our minds contain knowledge at the level of detail proposed by evolutionary psychologists. Scientists like Steven Pinker will make sure that evolutionary-psychological perspectives are injected into public debate for some time to come, though, and so it is important to grasp how freedom fits into this world view.

Pinker contributed two Darwinian cents' worth to the deluge of comment on President Clinton's sexual incontinence. Writing in the *New Yorker*, he observed that 'powerful male politicians may face temptations that most of their constituents do not'. In the past, rulers acknowledged their reproductive motives explicitly. Before the development of the state, no hominine male could ever exert sufficient power to monopolize a large number of females. The state enabled rulers to leap across primate social space, establishing harems for the first time in hominine history. That was before the notion of checks and balances, not to mention special investigators.

Since their reproductive interests tend to encourage females to be careful in the choice of mates, evolutionary psychologists expect women to be more averse to risk than men. They also expect men to take risks in pursuit of sexual opportunities. After all, males can succeed reproductively without living long enough to raise their children.

It is worth noting that the concept of risk being used here is more appropriate to Indiana Jones than to life as it is now lived in the developed world. Kingsley Browne argues that men get paid more than women because men are readier to take entrepreneurial, career or physical risks. Corporate raiding and rally driving may be comparable in spirit to village raiding and chasing large animals, but people in modern occupations face risks that are equally serious, if less glamorous. Women are disproportionately represented in occupations where they are responsible for the safety and well-being of others. If they fail in their duty to the children, the sick, the old or the disabled people for whom they care, they may face legal action and disciplinary procedures that could result in the loss of their livelihoods. Even in jobs that do not involve matters of life and death, the increasingly female workforce is more subject to unforgiving management regimes, under which minor lapses in performance can severely impair prospects in a competitive labour market.

Meanwhile, men in the United States' armed forces are resisting the onward march of women towards combat roles. Kingsley Browne considers that the male reaction has primordial roots, in the bonds that formed between male hunters on the notional savannah, and suggests that it is too viscerally strong to be overcome without weakening military efficiency. The remarkable thing about male bonds, from his account, is that although they form such a formidable obstacle to women in the military, they are apparently so negligible in the boardroom that they need not be mentioned in discussions of the glass ceiling. According to Browne, the glass ceiling is an illusion created by women's own psychological dispositions – weaker than men's towards status, and stronger towards family. He overlooks the possibility that men use their coalitionary tendencies to form mutually supporting boys' clubs, which exclude women from the informal networks that are vital to advancement. Browne's essay is not without its uses, though. It serves to confirm that some evolutionary psychology is just what lay sceptics fear it to be.

The importance of combat efficiency for American service personnel is open to question anyway, given the US military's extreme aversion to risking casualties. The US armed forces are not only the most powerful in the history of human conflict, but also the most risk-averse. In that respect, it was not entirely inappropriate for the US armed forces to have as their commander-in-chief a man who earlier in life had shown a reluctance to fight in Vietnam. That did not fit the evolutionary-psychological stereotype, but Bill Clinton certainly took other kinds of risks. As Steven Pinker has pointed out, this was only to be expected: 'Anybody who has what it takes to rise to the top of a profession – say, getting elected president – is likely to be a risktaker, a strategist and a moral utilitarian.'

If Pinker's comments had been made by an ordinary pundit, they would have passed unremarked. He is hardly the first person to observe that politicians are devious, after all. But Pinker is a Darwinian, which provoked the British journalist Polly Toynbee into a bizarre paraphrase: 'In other words, Clinton can't help it . . . None of us can help anything. First the Marxists, then the Freudians and now the Darwinians find a determinist answer to everything that makes us human.'[18]

Those certainly are 'other words'. Not a shred of what Toynbee quoted, directly or indirectly, implied that people cannot help their actions. If anything, Pinker's reference to risk, strategy and moral flexibility pointed to the range of choices that humans can make. And the presence of temptation does not mean that an individual is obliged to give in to it. As any Sunday school teacher will tell you, the whole point is that you have a choice.

Toynbee's outburst can be read as a complaint that the idea of choice had been denied by the two great secular belief systems of the nineteenth century – Marxism and Freudianism – and now the third system was reviving the tradition at the end of the twentieth century. It is true that all three systems pose profound questions about free will, and that crude versions of them give shallow answers. While the hotbed of contemporary biological determinism is behaviour genetics, not evolutionary psychology, the two disciplines have natural affinities. As each grows more confident, they are likely to grow together.

Just as mainstream Christianity has reached an accommodation with the basic principle of evolution, however, it is possible to reconcile evolutionary psychology with Christian moral thought, and the secularized versions of it which continue to provide the moral framework for a more agnostic society. If we accept that the mind is a combination of dedicated systems and general faculties that can reflect upon them, we have an architecture wholly compatible with the Christian model of the moral faculties. Christianity teaches that we have base impulses, but we have a conscience which can control them. In the terms of evolutionary psychology, the baser instincts are equivalent to the outputs of specialized modules, which produce rapid reactions along lines that are adaptive, or were originally. Conscience translates into general intelligence or a metarepresentational module, but whatever it's called, it does the same job.

Since Christianity has implanted a binary model of instinct and reflection so deeply in our framework for understanding the mind, it might seem peculiar that lay commentators do not place the idea of evolved psychological adaptations into this framework. Their perceptions arise, however, from a tradition of social science that has spent a century setting its face ever more firmly against biology, and the past half century with its convictions confirmed by the experience of Nazism. They cannot write the word 'Darwinism' without putting 'social' in front of it, and they cannot read the word 'biology' without hearing the word 'determinism'. It doesn't matter whether a particular manifestation of evolutionary thought contains any trace of the bogeys. These are articles of faith.

A lot of things just happen.

Stephen Jay Gould's book *Wonderful Life* describes a tranche of fossils, from a deposit known as the Burgess Shale, representing a range of forms that disappeared without trace during one of the great extinctions that have visited the Earth from time to time. There is nothing about the vanished forms to suggest that they were inferior in design to forms which survived. The moral that Gould draws is that life lies under chance's sway.

For people without religious faith, this can be a paradoxically reassuring message. Although it underlines the suspicion that the universe is capricious and merciless, it diminishes the unbelievers' sense of uncertainty, by affirming what they already suspected to be true. It appeals to a widespread secular mood of ontological unease, and agrees with it. The trope is less common in popular science than in science fiction; such as Douglas Adams's comedy *The Hitch-hiker's Guide to the Galaxy*, in which the central character finds spectacular confirmation for the nagging feeling he has always had, that there is something fundamentally wrong with the universe.

At this deep level, Gould's story resonates with truth. As an argument within evolutionary theory, it is less convincing. Gould presents his vision of contingency as a counter to the 'ultra-Darwinian' view that adaptation is all. 'Ultras', such as Richard Dawkins, deny that they underestimate the power of comets or other events to bring the sky down on heads that took millions of years to evolve. In *Unweaving the Rainbow*, Dawkins notes that questioners in American lecture audiences like to ask him about mass extinctions.[19] What he initially found puzzling about these inquiries was not the topic, but the challenging tone in which the questions were delivered. Then he realized that, having derived their ideas from Gould, the questioners were under the impression that mass extinctions demolish the gradualistic 'ultra-Darwinian' model of evolution. They do nothing of the kind. Comets and similar catastrophes are like storms which bring down trees in a forest, allowing flowers to spring up in the space cleared. The terrain cleared may be much larger, but the principle is the same; as are the principles of natural selection which then get to work in the new environment.

The difficulty with interpreting the fitness of the Burgess Shale organisms is that we know next to nothing of the environments they inhabited, and so we simply have no idea whether some were better adapted than others. Our knowledge of ancestral hominid environments is almost photographic by comparison.

A lot of stuff does just happen. This must be true at the level of invisible detail, as well as at the level of the comets. It must also be true that in a complex system like the mind, new effects will emerge from the interaction of systems that have evolved for other reasons. There will also be phenomena best explained by considerations other than selective ones. Gould's favourite bodily example is that of male nipples. Either there is some adaptive purpose, or they exist as a result of a developmental process common to both sexes, in which sexual characteristics are not settled until relatively late. Given the choice between these explanations, it is a waste of time to look for an adaptive story.

Evolutionary psychology, by contrast, has established itself on the presumption of adaptation. The search for the hand of selection has created a method, a research programme, and a distinctive group identity. How far this can take it is open to question. There is scarcely a shortage of psychological traits that can plausibly be suggested as adaptations. Evolutionary psychologists will not run out of ideas for the foreseeable future. The returns on the ideas may diminish, however.

As among the creatures of the Burgess Shale, some scientific papers flourish while others, although original and interesting, disappear without trace. Robert Trivers has noted a paper by

W. D. Hamilton and M. Zuk, arguing that sexual reproduction is a strategy for reshuffling genes to keep ahead of parasites, as an example of how a scientific organism can flourish in the right niche.[20] According to Trivers, this paper is cited everywhere, which shows the advantages of basing a discipline around a single paradigm. It brings ideas into close contact, encouraging them to replicate rapidly.

This is a good argument for working within the paradigm. The danger, however, is that the discipline becomes exclusive, failing to accept the need for outside input. Schools of thought that follow such a route may increase their output, and consolidate their institutional bases. But they may also end up with a narcissistic body of inward-looking ideas, trading endless mutual references. Although Darwinians may deny that this could happen to them, as they are scientists and that kind of thing only happens in the post-modernist humanities, less partisan observers may be less sanguine.

Even at this stage, it is not clear that the pioneers of evolutionary psychology will maintain their early momentum. Some of the next steps in the study of the evolved mind may be made by scholars who adhere to the paradigm, but do not see themselves as part of the movement. Researchers with backgrounds in anthropology or archaeology may be less concerned to flush out adaptations, and readier to accept that in human prehistory, sometimes a quirk is just a quirk.

It's time to get off the savannah.

'Our modern skulls house a stone age mind,' according to Leda Cosmides and John Tooby.[21] This formula has become one of evolutionary psychology's most successful sound bites. It speaks to the idea that many of our discontents arise from a mismatch between the environment that made us and the ones we have made for ourselves. In the EEA, or environment of evolutionary adaptedness, we have a three-letter acronym for Arcadia.

As both sympathetic and critical commentators have noted, this axiom of evolutionary psychology has been endorsed by Theodore Kaczynski, the Unabomber, who conducted a terrorist campaign in the United States against targets he considered part of a repressive technoscientific order.[22] In his 'Manifesto', the Unabomber attributed 'the social and psychological problems of modern society to the fact that society requires people to live under conditions radically different from those under which the human race evolved and to behave in ways that conflict with the patterns of behavior that the human race developed while living under the earlier conditions'.[23] He is scarcely alone in his suspicions, which are shared by many who feel that we must find a way of living that is in harmony with nature, for its sake as well as ours. This was the take-home message of the twentieth century.

Despite his analysis, the Unabomber went to great lengths to keep clear of his evolutionary roots. Lest he be used to imply guilt by association, it should be noted that Kaczynski chose to live under conditions radically different from those in which humankind evolved. Bombs were not part of the ancestral environment, nor was the postal service he used to deliver them. Nor, more fundamentally, was solitude. The ancestral human condition was one of group living, in which social relationships were intense, complex and dynamic. Kaczynski lived a quintessentially modern life of alienated isolation, but it had the semblance of tradition because it was lived in a rural cabin rather than a city apartment.

Like the Unabomber's way of life, ideas about the relationship between nature and artifice are often more complicated, and contradictory, than they at first appear. The idea of stone age minds in modern skulls may be a gift for a magazine editor seeking an evolutionary bullet point. For evolutionary theorists, it may prove less satisfactory in the long run.

Stone age minds have helped popularize evolutionary psychology not just because the idea is easy to grasp, but because it has finessed a serious political difficulty for the emerging school of thought. By defining the project as a search for the evolved universals of human nature, evolutionary psychologists distanced themselves from efforts to explain differences between modern populations in biological terms. References to an undifferentiated 'savannah' underscored the idea of fundamental unity. We are one because we evolved in one place.

This is something of a fiction. In *How The Mind Works*, Steven Pinker explains that humans are adapted to two environments, the African savannah and the rest of the world. The title of the section is 'The Suburban Savanna', but hominids have been in the rest of the world for a long time. The lowest estimates arise from the 'Out of Africa' theory, which proposes that all living humans are descended from a single population which began to disperse out of Africa about 100,000 years ago.

Since humans did not develop a recognizable symbolic culture until the last 50,000 years, and all of them lived by hunting and gathering until around 10,000 years ago, many of them spent tens of thousands of years at least in environments that maintained the old selective pressures, but were not in Africa. If the selective pressures acting on the mind were the same everywhere – in other words, that they arose from group living, rather than geography or other species – then human nature should have remained universal. Nevertheless, the possibility of populations evolving separately is the last thing that evolutionary psychology wants to contemplate, politically speaking.

This is one of the reasons why, according to the anthropologist William Irons, criticism of the EEA concept has come not from those who define themselves as evolutionary psychologists in the narrow sense, but from scholars variously labelled behavioural ecologists, evolutionary ecologists, human palaeontologists, or just plain sociobiologists.[24] One important aspect of the evolutionary psychologists' vision that the critics have questioned is the role attributed to hunting and gathering.

In his original formulation of the EEA concept, John Bowlby

contrasted two million years of hunting and gathering with a few thousand recent years in which humans have developed other ways of life. Jerome Barkow, Leda Cosmides and John Tooby echoed him in *The Adapted Mind*, a collective manifesto for evolutionary psychology. 'The few thousand years since the scattered appearance of agriculture is only a small stretch in evolutionary terms,' they observed, 'less than one per cent of the two million years our ancestors spent as Pleistocene hunter-gatherers.'[25]

Lumping two million years of hominid foraging strategies together does not show a great deal of respect for historical detail, however. One stone tool may look much like another to some evolutionary psychologists, but as Irons points out, a lot of evolution has taken place among hominids over the past two million years. The several species that came and went during that era had more than one way of life.

Another critic of the EEA, Robert Foley, points out that modern human hunter-gatherers do not share a single way of life either.[26] Group sizes range from nine to 1,500. They may move as often as once a week, covering up to 3,600km in a year, or they may stay put. Men may provide all the food eaten by the group, or as little as a fifth of it. Different populations may behave quite differently in similar landscapes. Central Australia and the Kalahari are both arid, but the aboriginal Australians have a social structure that is the toast of social anthropology for its complexity, while the San of the Kalahari take a relaxed and informal approach by comparison. Hunter-gatherers thus have wide options even in exacting habitats – to which the last of them are mostly confined. When all people were foragers, on benign as well as meagre lands, they must have been still more diverse than the modern peoples who follow a similar way of life.

Instead of a unitary environment of evolutionary adapt-edness, underpinning a uniform hunter-gatherer way of life, Foley proposes that the human evolutionary heritage should be seen as a combination of deep primate roots, and far more shallow features related to the more recent past of modern humans. Traits such as sociality are very ancient, and may be partly under genetic control. As the modern hunter-gatherer way of life emerged while modern humans were dispersing into different parts of the world, however, its connection with the common ancestral environment is weak.

William Irons' critique of what he calls 'Pleistocentrism' is different in intent. Not only does he believe in a multi-tude of specialized adaptations, but he also argues that in many cases their adaptive effects may have survived into historical times. He suggests that they may guide behav-iour such as status striving, incest avoidance, and maternal bonding.

When a mother and her newly born baby are separated by a bureaucratic mistake in a hospital, the incident is loudly deplored in the media. Popular opinion has incor-porated the idea that if mother and child are not together in the period immediately following birth, the opportunity for a special bond between them will be missed, possibly leading to serious consequences later in the child's life. In the early 1970s, medical professionals were impressed by studies which indicated that something special happened between mother and child in the wake of birth, and bad things would happen between them if it did not. Medical opinion was subsequently swayed in the opposite direc-tion by a round of studies which criticized the original findings. In the words of Martin Daly and Margo Wilson, this turnaround threw the baby out with the bathwater.[27]

One of the things that contributed to the backlash against

bonding was the fact that some mothers were at first indifferent to their infants, even when hospital policy encouraged them to bond. Daly and Wilson claim this observation for an evolutionary model by suggesting that the aftermath of birth is a moment in which to take stock. If all is well, with both the child and its discernible prospects, then its birth can indeed be the occasion of love and celebration. If not, the mother may decide to abandon the baby. In traditional societies, that means killing it.

Daly and Wilson's reflections on maternal bonding are part of a longer discussion of infanticide, which they introduce with a description of the birth ritual of the Ayoreo, who live in the borderland of Bolivia and Paraguay. No culture could devise a practice which more starkly dramatized the treatment of birth as a moment of choice. In the Ayoreo language, the word for 'to be born' is also the word for 'to fall'. A woman gives birth in the forest, supported by the branches of trees, and surrounded by close kinswomen. Her companions moisten a patch of earth under the branch, to soften it, and dig a hole nearby. When the baby falls to earth, they examine it. If it is not deemed to be sound, the infant is pushed into the hole with a stick, and is buried without ever being touched by a human hand.

Ayoreo women are not callous. Burying a baby is experienced as a tragic event, and women who have done it find it painful to talk about. (Zu/'hoasi women are reported to have a similar attitude to infanticide.) Although Daly and Wilson do not refer to the Ayoreo in their comments on bonding, the implication seems to be that the moment after birth is no more than a small window in which affection can be suspended. The countless women who experience love for their babies unconditionally and immediately, without a pause, suggests that the suspension is anything but automatic.

It also underlines the fact that hospital births can be successful and fulfilling experiences, despite the disparaging terms in which William Irons refers to modern medical services. Against these he contrasts an image of ancestral childbirth that is simultaneously idyllic and brutal. 'It is important to note that in many traditional societies, women gave birth in a secluded location with only a few female kin present,' he says, but offers no supporting examples. Nor do Daly and Wilson, apart from the Ayoreo. Irons defines the adaptively relevant environment for maternal bonding as one which permits women to take the decision of life or death for their infants 'with the support of close kin and without the interference of unrelated individuals'. Environments such as these persisted until modern health services replaced kinfolk with an 'impersonal bureaucracy', a description which thousands of nurses, midwives and doctors could justifiably consider an insult.

Although Irons acknowledges that the possibility of adaptive infanticide raises a large moral issue, he immediately balances it by suggesting that the 'possibility that hospital procedures can raise the risk of child abuse is another serious moral issue'. The implication of moral equivalence between ancestral baby-killing practices and modern healthcare rounds off a reactionary subtext whose opposition between ancestral and modern is as artificial as that of the EEA.

The vital insight which the EEA concept crystallized was that evolutionary perspectives do not require us to assume that modern human behaviour is adaptive. That was a fundamental precondition for a real understanding of the role biology plays in human affairs. It helped define a peculiarly human relationship to nature, by pointing to the complications that intrude between human nature and the purposes that human nature originally evolved to serve.

In this way it affirmed human possibilities – not necessarily productive or desirable possibilities, just complex ones that confound pat accounts of the human condition. Concepts of adaptive environments become obstructive, however, when they oversimplify evolutionary history, or collude with legends of the Fall.

*

All the same, we are still not happy. We are oppressed by crowds, alienated by the impersonality of human interaction in cities, uncomfortable with social networks that allow intimate friends to meet only after careful negotiation and several weeks' notice, guilty about the treatment of elderly kin. We are chronically anxious to improve our standard of living, even if it is more than adequate already. Under conditions of everlasting agitation and uncertainty, we are nagged by the feeling that we have left some important things behind, but we are not sure exactly what they are.

Given our discontents, we should not leave the idea of ancestral adaptation without at least asking whether we have roots with which we should get back in touch. Robert Wright, author of *The Moral Animal*, has suggested that by reconstructing ancestral conditions, evolutionary psychology can help make sense of modern unhappinesses. Among these he includes the depression of the housewife, alone with her children in an anonymous suburb, or of the working woman who will 'hand her child over to someone she barely knows and then head off for ten hours of work'. The popularity of the 1980s bar-room sitcom *Cheers*, he cites Leda Cosmides and John Tooby as observing, may reflect a 'visceral yearning for the world of our ancestors', in which acquaintances encountered each other regularly and randomly, without need for scheduling; 'where there were spats and rivalries, yes, but where grievances were heard in short order and tensions thus resolved'.[28]

Such visceral yearnings could also be invoked to explain less innocuous modern Western phenomena, such as road rage attacks. After all, if a grievance is held against a stranger who may be out of reach and miles away within minutes, then retribution must be instant and emphatic in order to be reliable. Nor can it be assumed that ancestral spats and rivalries were typically resolved with the equivalent of hugs and rounds of beers. To judge by contemporary societies, a life of hunting and gathering is not typically one of peace among men. Even among the Zu/'hoasi (!Kung), generally characterized as a pacific people, twenty-two homicides were recorded over a twenty-five-year period. Fifteen of these occurred in the course of blood feuds, illustrating that tensions may be perpetuated rather than swiftly resolved in small traditional groups.[29]

The twenty-two killings correspond to a rate of 293 homicides per million people per year. Although the reputation of the Zu/'hoasi has survived data such as these, southern Africa as a whole has become associated with violence, thanks not just to wars but the frightful crime levels experienced in South Africa. In the first quarter of 1998, murder rates ranged from 80 per million, in the rural north-east, to 680 per million, in Cape Town. The Zu/'hoasi rate thus fell in the middle of the South African range, and was more than twenty times higher than the homicide rate recorded in England and Wales in 1996, of thirteen per million.[30] Although many Westerners might feel that hunting and gathering is a nice place to visit, they wouldn't live there if you paid them.

The trouble with traditional societies as guides to where we took wrong turnings is that they so readily remind us why we moved away in the first place. If we want to invoke tradition, though, we would do well to bear them in mind. Towards the end of the twentieth century, revisionist sexual politics made a cliché of the notion that masculinity was in crisis because industrial changes had robbed men of their traditional roles as breadwinners, which women had now assumed. The tradition that men go off to do a day's work or a shift, while women stay at home, is relatively recent and largely Western. In other societies, including hunter-gatherer ones, the traditional male role has been to spend as much of the day as possible in the company of other men, chatting and engaging in competitive diversions, while women look after children and produce or gather food. By and large, men appear to treat work as an obstacle rather than a means to the development of identity.

Anthropology and biologically based sciences of behaviour would agree that men suffer when they lose status, but these disciplines have concentrated on what happens when male status changes with respect to other males. They have not had to consider what happens when males lose status with respect to females. We are into a new game here, and there will often be no obvious evolutionary lesson on which to draw.

In many, if not most cases, none will be necessary. 'We aren't designed to stand on crowded subway platforms, or to live in suburbs next door to people we never talk to, or to get hired or fired, or to watch the evening news,' says Robert Wright. 'This disjunction between the contexts of our design and of our lives is probably responsible for much psychopathology, as well as much suffering of a less dramatic sort.'[31] Maybe so, but arguments in this line are too close for their own good to the declaration that if God had meant us to fly, he would have given us wings. It is easy to select a handful of modern experiences one finds disagreeable, and blame the disagreeableness on a lack of adaptation. We are no more adapted to standing in crowded sports stadiums than to subway platforms, though, and yet hundreds of thousands of us pay millions for the privilege. We may spend much of our days talking to the same small group of individuals, but at our places of work rather than our homes. Our ancestors did not get hired or fired, but their fortunes (not to mention their lifespans) undoubtedly depended upon their ability to achieve status and maintain their places within their groups. The evening news is neither here nor there; which is why its impact is generally shallow.

As for flying, it is true that fear of it is widespread. If evolution had meant us to fly, it would have selected against a fear of heights, which otherwise would be adaptive. But the remarkable thing about flying is how readily most people take to it, whether they are tourists describing the holiday arc from northern Europe to Mediterranean resorts, or peasants huddled over their sacks of produce in military turboprops, weaving among Andean peaks.

An instinctive aversion to heights could underlie some people's fear of flying; though there seem to be many people who don't mind heights, but are afraid of flying, and vice versa. But the more powerful adaptation in this case would appear to be behavioural flexibility – as it would appear to be in all other off-savannah human behaviours, by virtue of the fact that they exist.

We are back up against the difficulty that evolutionary theory cannot say how strong our instincts are. Whether you prefer to conceive of the mind as a mass of specialized circuits, or allow a greater role for general mechanisms, the relative strengths of the systems cannot be measured. It may be possible to deduce that a particular adaptation evolved, and how it worked in ancestral environments, but there is no clear path from there to here to show us how we could be happier and healthier.

Some initial steps in tracing such a path have been made by the scientist on whose work modern Darwinism is founded. In collaboration with Randolph M. Nesse, a psychiatrist, George Williams has sketched out a project of Darwinian medicine, which would consider symptoms in the light of their probable evolutionary background.[32] It stands to win sympathy from quarters generally unsympathetic to Darwinism, since it is inclined to question whether modern medicines are preferable to letting nature take its course. Fever appears to be not the work of a pathogen, but the body's evolved response to challenges from pathogens. This implies that if aspirin is used to reduce fever and its associated discomfort, the pressure on the pathogens may be eased. In support of this hypothesis, Darwinians cite a study which found that chickenpox lasted longer in children who were treated with acetaminophen, which lowers fever, than among those given a placebo.[33]

For many patients, especially children and the elderly, more may be at stake than an extra few days' illness. A Darwinian physician would have to steer between the Scylla of dangerously elevated temperatures, which might lead to delirium or seizures, and the Charybdis of weakened defences against pathogens. Fever may have been a heroic adaptation, which cured or killed. Now it is one more thing we are obliged to manage.

The same is true of emotional trouble, which Nesse and Williams also wish to bring under the Darwinian umbrella. 'Unpleasant emotions can be thought of as defences akin to pain and vomiting,' Nesse has said. 'Just as the capacity for physical pain has evolved to protect us from immediate and future tissue damage, the capacity for anxiety has evolved to protect us against future dangers and other kinds of threats. Just as the capacity for experiencing fatigue has evolved to protect us from over-exertion, the capacity for sadness may have evolved to prevent additional losses.'[34]

But what if anxiety or sadness become the background colour of a person's life? A bout of vomiting is tolerable, and so is a moment of anxiety. Vomiting that persisted day after day would be unendurable, and the effects of chronic anxiety may be similarly debilitating. Implicit in both Nesse's analogy and other portrayals of the ancestral condition is the assumption that disorders used to be acute rather than chronic. Either the infection or the injury killed you, or you survived, but either way matters were swiftly resolved. The same is taken to be true for emotional afflictions. 'What isn't natural is going crazy,' declares Robert Wright. Sadness protracted into depression and anxiety sustained chronically are 'largely diseases of modernity'. He refers to a study in rural Samoa which found levels of cortisol, a marker of stress, that were extremely low by Western standards, and a study of depression among a New Guinean people which recorded an incidence of zero.[35]

This does not mean that Samoans and New Guineans are happy, of course. Each society may be unhappy in its own way. Nor can it be assumed that people in traditional or ancestral societies were spared long periods of unhappiness. There is good anthropological reason to suppose that infanticide has been wide-spread in traditional societies, and no good reason to suppose that women in these societies who kill their infants are spared the grief that afflicts bereaved mothers everywhere. Since their acts are sanctioned by custom, if not ordained by it, their reluctance to talk about infanticide seems more likely to stem from grief rather than guilt. Since this reluctance persists indefinitely, it seems likely that the pain does too. The high rates of infant mortality observed among contemporary foragers underlines the likelihood that the loss of children was a common experience in traditional and ancestral times. If mothers caused some of these bereavements deliberately, how would their sorrow help them? According to Randolph Nesse, sadness could have the adaptive effect of impelling people to take steps which would reduce the risk of further loss. Yet evolutionary thinkers such as William Irons must assume that these strategic choices were adaptive. It might be in a woman's reproductive interests to kill or abandon a child more than once in her life. If so, her sadness over a previous loss would be maladaptive.

The point of harping on this doleful theme is to indicate the kind of complexities such a perspective would have to resolve, and to raise a counterbalance against the assumption that people in ancestral times either bounced back or were dead. It is not to deny that an evolutionary perspective on the emotions is worth developing. The management of emotion is becoming one of medicine's top priorities. One way of assessing these priorities is to calculate the number of disability-adjusted life years, or DALYs, associated with different illnesses and hazards. Somebody who is killed in an accident at the age of fifty, instead of living to eighty, has lost thirty years. If the accident leaves them with 50 per cent disability, the adjustment calculation deems them to have lost fifteen years. Reckoning World Health Organisation figures in terms of DALYs, the top three medical conditions are lower respiratory infections, diarrhoea and perinatal problems.

The fourth is depression, right behind the familiar curses of world poverty, and ahead of heart disease, tuberculosis, strokes, cancer or any other of the afflictions people most fear. Extrapolated, the trends indicate that ischaemic heart disease will top the league table by 2020, and depression will be in second place.[36]

The treatment of depression has been revolutionized by the class of drugs known as selective serotonin reuptake inhibitors, or SSRIs. More than twenty million Americans take them; the name of the leading brand, Prozac, has become a household word. Although it would appear that the depression market is anything but saturated, the SSRIs are being promoted as more than anti-depressants. Fluoxetine, under its alias of Prozac, has achieved the kind of profile among pharmaceuticals that the Windows operating system enjoys in the world of computers. Like Windows, its performance advantages over competing products are not so clear as to account for its commanding lead, which must owe much to brand recognition. The dominance of fluoxetine in the core anti-depressant market may be one reason why a competitor, paroxetine, was relaunched as a cure for social anxiety disorder, an extreme form of shyness said by the American Psychiatric Association to afflict one person in eight.[37]

That could just be the beginning. The SSRIs offer the prospect of feeling 'better than well'. Just feeling OK will no longer do. If people live by the expectation that their standard of living and the standard of their consumer technology should follow an upward course indefinitely, they will find it easy to form similar expectations about their levels of happiness and their pharmaceutical technology. They will also feel compelled to improve their states of mind through fear of competition. When you see a colleague setting to work in the morning, relaxed and cheerful while others are grey, or an acquaintance holding court in the evening, at the centre of a laughing and attentive circle, you will suspect a case of 'better than well'. This is not simply a question of envy, but of whether your associate is using performance-enhancing drugs to gain advantage in competition for professional advancement, social status, or sexual opportunities. Can you afford not to feel as good as your competitors?

Randolph Nesse does not doubt that these drugs work. Their progress has been dazzling, and he foresees a succession of new drugs acting upon increasingly specific emotional zones. He cautions, however, that progress in developing new pharmaceuticals has not been matched by progress in understanding how SSRIs actually achieve their effects. Even basic questions remain unresolved, such as why people typically do not enjoy the benefits of SSRIs until several weeks into the course of treatment, although serotonin levels rise the day after they start taking the drugs.[38]

Nesse's deeper concern is with the ethical questions raised by a burgeoning emotional pharmacopoeia. Faced with a child in distress from fever, the obvious humane response is to give aspirin. Darwinian considerations, and the chickenpox study, suggest that the obvious response may not be the best one for the patient. But although the decision in an individual case may be tricky, the episode will be short and its end easy to recognize. Also, comparing different treatments is the bread and butter of clinical research, so generating data to inform clinical practice should be relatively straightforward. When distress is emotional rather than physical, deciding on the best response may require moral intuition rather than clinical experience. If somebody is suffering because they have been bereaved, is it more humane to relieve their misery with drugs, or to let their grieving take its course? If medication is considered appropriate, should it commence after an interval of months, or weeks, or days?

By improving the prospects of success and reducing the drawbacks, the SSRIs have made prescribing anti-depressants much easier. As new drugs are launched, with promises of fewer side-effects than ever, prescription will become the default option. Nesse predicts that the bereaved will be urged by their doctors to medicate their grief within a fortnight. SSRIs will also be used for a wider range of complaints, and their successors will be promoted as remedies for any emotional ill and discontent that can be named.

In the present state of its art, evolutionary medicine can only be used to sound cautionary notes. There could be no greater challenge for applied Darwinism in the coming century than to develop practical therapeutic strategies which adequately represent the relationship between the evolved basis of psychological traits and their operation in modern environments. At this stage, it is not clear whether such a project could ever be feasible. If not, people will have to rely on folk wisdom. Some will decide that nature knows best. Others will decide that if God had meant us to be unhappy, he wouldn't have given us Prozac.

There is no such thing as a left-handed screwdriver.

In the old days, when men who worked in factories made their own amusements, youths on their first days as apprentices would traditionally be sent to the stores to ask for a long weight or a left-handed screwdriver. They learned that 'weight' was spelt 'wait' in this case, and that screwdrivers could be used in either hand. Like screwdrivers, evolutionary tools can be used whether one's preference is for the Right or the Left.

The distinction is not as clear as it used to be, but the Left can still be recognized in the belief that cohesive societies are healthy societies. A left-wing disposition tends to find virtue in collective arrangements, which it sees as the means by which cohesion is achieved. Some of these, such as collective bargaining by trade unions, are left-wing in the traditional sense. Others, such as the social market traditions of continental Europe, or Britain's National Health Service, also command extensive support from the Right. Much of the Right also agrees with the Left on the importance of civil society, the organic networks of association that are as essential as the structures of the state, although they are sometimes hard to see; just as the microscopic structures of a cell's cytoplasm are vital, although less noticeable than the nucleus. Where Left and Right do diverge, however, is on the question of equality. The Right asserts the need for equality of opportunity, but sees no harm in inequality of outcome; indeed, may positively welcome it. The Left is uncomfortable with inequality of outcome, seeing it as a threat to social cohesion.

Evolutionary theory, like God before it, might be suspected of being a Conservative. 'Let our children grow tall and some taller than others if they have it in them to do so,' proclaimed Margaret Thatcher. What ideological doctrine could be more in keeping with the laws of variation and natural selection? And what scientific paradigm could possibly appeal to a right-wing world view more than modern Darwinism, founded on a rejection of the group in favour of the individual?

It is true that modern Darwinism attracts some extremely right-wing individuals, maintaining exactly the tradition it has tried to shake off. One e-mail discussion group, associated with a prominent academic association devoted to human evolutionary studies, collapsed after a contributor terminated a thread of argument by posting long tracts denying the truth about the Nazi extermination of the Jews. Although the Holocaust was declared off limits for its successor, the reconstituted group remained a forum for opinions that would make anything from the housetrained Right look weakly pink.

More significantly, it is also true that because evolutionary psychology is behaviour genetics' fellow traveller, scholars of otherwise conventional political views may entertain notions about inherited differences that would offend lay people on both the Left and the Right of the political centre-ground. But there is no necessary association between evolutionary theory and the Right. John Maynard Smith has noted that the thinking public tends to associate the Right with reductionism, and the Left with holistic approaches to the study of how organisms develop. Yet although he belongs to the reductionist camp, he has never voted for Margaret Thatcher or her successors. His Left credentials are more clearly indicated by the period he spent in his youth, trying to reconcile his Marxism with his biology by attempting to demonstrate the inheritance of acquired characteristics. Although he himself became more reductionist as he became less Marxist, Maynard Smith also pointed out that this inverse relationship does not hold in the economic sciences. Advocates of holistic biology emphasize the capacity of complex systems to organize themselves. In economics, this corresponds to what Adam Smith called the 'invisible hand', which supposedly distributes the necessaries of life among rich and poor alike.[39]

One or two others, notably Chris Knight, have moved into human evolutionary theory from a Marxist background. Another individual who ended up in evolutionary circles after a formative period as a radical, perhaps not of the Left but implacably opposed to the Right, was Huey P. Newton. In 1966, Newton and Bobby Seale founded the Black Panthers, an armed and militant organization that challenged the American police, with predictably severe consequences for its cadres. Newton transformed the party's strategy from one of armed resistance to providing social services for black people in 1971, but spent much of the decade with a murder charge over his head. After two hung juries, he obtained a doctorate and worked with Robert Trivers. They collaborated on a project about self-deception, in which they studied recordings of conversations in the cockpit of an airliner that took off in icy conditions and crashed into the Potomac river: one of their colleagues was among the passengers killed. Newton himself was killed in 1989, shot dead on a street in Oakland, California. Robert Trivers observed that Newton was suited to the study of deception, being a proficient con-man himself. He also described Newton as 'a genuine political radical . . . it never occurred to him that evolution would put him in a corner he couldn't manage'.[40]

Newton and Seale had originally raised the Panther banner in the name of self-defence; 'fighting the power', as a later black radical slogan was to put it. Anybody who wishes to fight power must understand where it comes from, 'who has it, how they get it, how it is used, and what are its consequences', in the words of Barbara Smuts, who identifies herself as a feminist and an evolutionary biologist. Anybody who wants their understanding of power to start from first principles must start from evolutionary theory. Nowhere is this truer than in feminism, for as Smuts points out in her article 'The evolutionary origins of patriarchy', feminist theory and evolutionary theory share the same fundamental concerns. Both are about sex and power.[41]

Evolutionary theory, according to Smuts, answers a question that feminist theory does not reach far enough to ask. Feminist arguments take it as given that men want to control women; an evolutionary perspective gives the reason why they want to do so. They are so successful at it, Smuts argues, because several reproductive strategies used by male primates have become especially elaborated in our species. By primate standards, human males and females are unusually unequal; but for Smuts, the point of primatological comparison is to redress the imbalance. Each trait suggests a counter-strategy. The characteristic pattern of coalitions is that those between males are strong, while those between females are relatively weak; so women need strong political solidarity, with the goal of creating 'strong institutionalized protection of women from male violence and other forms of domination'. Men's power to control women has increased with their power to control resources; so women need property rights and opportunities to develop economic resources of their own.

Males also control women more effectively when they dominate other males. If other males are able to intervene when a male primate tries to coerce a female into mating – with their own reproductive opportunities as an incentive – sexual coercion is unlikely to emerge as an effective strategy. It is not a coincidence that the Moroccan potentate Moulay Ismail had a reproductive score of 888, and that he was known as 'the Bloodthirsty'. Equality between women and men requires greater equality among men.

Smuts succeeds in matching a conventional evolutionary analysis to a familiar feminist agenda, without bending one to fit the other. Her article is a legitimate political reading of science. Its coherence illustrates the coincidence of perspectives between feminism and Darwinism, two systems of thought with female choice at their centres.

There are evolutionary psychologists in the actually-existing, domain-specific, species-typical sense; and there is evolutionary psychology in a wider sense, of inquiry into human affairs that employs evolutionary and psychological perspectives. Dimensions such as these have become all the more important since the horizons of politics were narrowed by the general acceptance of orthodox market economics.

With politics reduced to a rump, economics and psychology are left to govern the vacated territory between them. Orthodox economics believes that market forces will induce individuals to behave in a basically rational fashion. Psychology understands individuals as emotional, subject to conflict and frequently irrational. Although these two perspectives are not necessarily contradictory, the difference between them becomes more obtrusive as they assume greater salience in the interpretation of human behaviour.

It is possible to reconcile the two without drawing upon evolutionary theory, but as the work of Robert H. Frank shows, Darwinism helps the synthesis cohere. Frank, an economist based at Cornell University in the US, handled Hamilton's kin selection and Trivers's reciprocal altruism like a natural in his book *Passions Within Reason*, a dextrous study of why it makes sense to be emotional.[42]

Signal theory is among the biological strands on which Frank draws. He points out that anger can be useful as an honest indicator of how an individual will react if his interests are infringed. The argument can be illustrated by the stereotypically contemporary example of an incident in which a motorist parks his car so as to block a driveway. The property owner emerges from the house and informs the motorist that he will break his legs if the car isn't moved. If the aggrieved party makes his threat in an even tone of voice, without adopting a threatening posture, he will probably be ignored. In itself, the threat is not credible because it is irrational. No sensible person is going to risk an almost certain prison sentence over an inconvenience that is probably trivial. If all that the aggrieved party is prepared to commit to the threat is the cheap signal of a few words, he will probably not be prepared to incur the heavy costs of delivering on it. But if the threat is accompanied by costly signals that are hard to fake, such as a red face and a posture indicating real anger, the threatened party is likely to decide that the probable consequences far outweigh the benefits of the parking space.

The property owner's response is particularly effective against his neighbours, the individuals most likely to inconvenience him. His reactions will make a lasting impression on them, especially as they are likely to observe a pattern of aggressive behaviour. He may not be popular, but he will enjoy unobstructed access to his garage. The moral is applicable to most settings, including ancestral ones: genuine emotion can make irrational behaviour result in better outcomes than a rational response.

Most of *Passions Within Reason* is not about anti-social behaviour, though. Scholars are more interested in what they find surprising than what they expect, and economists are not surprised when people behave selfishly. Frank's essay is a reflection upon why people are ready to behave unselfishly in so many ways, great or tiny. It celebrates the passer-by who returns a lost wallet, the bone marrow donor, the citizen who walks in the rain to the polling station, although a single vote is most unlikely to tip the result, the diner who leaves a tip on leaving a restaurant in a town he will probably never visit again, and the man who jumped into the icy Potomac river one night in January 1982, to rescue a passenger from the crashed airliner whose pilots' fatal psychology was subsequently investigated by Robert Trivers and Huey Newton.

Frank does not use examples like these to imply that we have risen above inclusive fitness and reciprocal altruism, or that evolutionary models are in some other way unworthy to describe the human moral condition. He is concerned to extend theories of human behaviour in ways that account for actions that are altruistic beyond the call of shared genes or reciprocity, and that make sense within an overall framework of self-interest, genetic and economic.

Many examples of apparent altruism are easy to discount against considerations of reciprocity or inclusive fitness. Tissue donors are often related to recipients. A diner who leaves a tip in a restaurant he may visit again is investing for good service in the future. Selfish motives are often suspected behind charity, as Thorstein Veblen observed. 'It is a matter of sufficient notoriety to have become a commonplace jest that extraneous motives are commonly present among the incentives to this class of work – motives of a self-regarding kind, and especially the motive of an invidious distinction,' he remarked in *The Theory of the Leisure Class*. The jest was still commonplace a century later, when characters played by the comic actors Harry Enfield and Paul Whitehouse made a catchphrase of the boast 'he does a lot of work for charity, but he doesn't like to talk about it'. Their joke was that the DJ thus praised was continually cajoling his partner into talking about it, even to the extent of holding up cue cards with the phrases written on them.

Despite this long and thriving tradition of scepticism, though, public displays of charity continue to thrive unabated. Charitable donations or works by people of high status are effective however their motivation is perceived, because the donors demonstrate their wealth by showing that they can afford to give some of it away. Amotz and Avishag Zahavi cite charitable donations as an example of their Handicap Principle, under which fitness is signalled by costly displays such as an unwieldy crown of antlers, or risky behaviour in the presence of predators. They draw out one of the connections between wealth and status by noting that charitable donations are often made by prominent people at public gatherings, putting pressure on others present to do likewise so as to avoid losing face.[43]

Specious examples of altruism make the real thing all the more impressive. One way to reconcile this shining residue with evolutionary theory would be to talk of memes, in the form of religious or other moral precepts that induce people to do the right thing even if it is not immediately good for them. Frank offers a different suggestion. In many of our activities and relationships, it is fundamentally important for us to demonstrate commitment. We need to convey to our partners that we will continue to act considerately towards them, even if circumstances change and it ceases to be in our interest to do so. If they are to feel confident about our commitment, they have to believe that our hearts are in it. If they read different messages about our emotions and our stated intentions, their suspicions will be aroused. Frank suggests that our commitment to doing the right thing in one relationship may be weakened if we do not maintain similar commitments in all our relationships. Leaving a tip in a restaurant you will never revisit is good moral exercise, as is paying your taxes in full while your acquaintances evade theirs. The more you do the right thing, the more you mean it.

On the other hand, the more you are immersed in systems based on self-interest, the more selfish you are likely to become. An unusual gaming genre, part play and part experiment, has developed from the mathematical models of co-operation and cheating pioneered by Robert Axelrod. The most successful strategy that emerged from Axelrod's computerized tournaments was known as 'Tit for Tat'. Each player begins by co-operating, and from then on matches the other's moves. This is known as 'Tit for Tat'. One player gives the other a token; the second gives a token in return, and so on. Each player benefits and the situation is stable, unless one player suddenly defects from the arrangement. The appropriate response for the other player is to do the same; and now both players are worse off than they were before. ('Tit for Tat' actually works better if a little leeway is allowed, and defections are not invariably punished.)

Although the games get more complex, and can include more than two players, they still turn on the balance between co-operation and defection. When researchers organized game sessions among different groups, a party of economics graduates proved significantly more likely to defect than any of the others. Speaking of our behaviour as a whole, Frank observes that 'the self-interest model, by encouraging us to expect the worst in others, does seem to have brought out the worst in us'.

Robert Frank's left-wing credentials extend only as far as voting Democrat. He describes himself as right of centre, and his belief in the market is orthodox. Yet he and his colleague Philip J. Cook have written a critique of contemporary markets more incisive than all but a few of the commentaries to emerge from left of centre in recent years.[44] *The Winner-Take-All Society* deals with the kind of market that has long been familiar in sport and entertainment, in which a handful of individuals at the top of their professions command a massive proportion of the market's wealth. Competitors in these markets become caught up in runaway bidding processes, because immense rewards turn on marginal differences in performance. A footballer does not command a price of five million pounds because he is five times as good as his colleague who fetches one million pounds. The club which buys his services is investing in the likelihood that he will perform marginally better than the other top players he faces, and therefore help his side secure the vast revenues that now accrue to the elite of football clubs.

The same logic applies to boardroom salaries. A company lures a new chief executive with the offer of half a million pounds a year because that is probably a price worth paying for an individual on whose performance the company's survival may depend – and who will go to a rival if not given lavish inducements. This is becoming the way of things in more and more sectors of the economy, as winner-take-all markets develop in areas such as publishing, the legal profession, journalism, medicine and the upper reaches of academe. In the winner-take-all markets of culture, a small number of musicians, actors and authors command earnings out of all proportion to their relative ability, while their material tends to become more formulaic, more sensational, or both. Frank and Cook sum up the process in the subtitle of their book: 'more and more Americans compete for ever fewer and bigger prizes, encouraging economic waste, income inequality, and an impoverished cultural life'.

If the money is concentrated at the top of a few sectors, Frank and Cook point out, these will attract large numbers of candidates, the vast majority of whom are doomed to fail by the very nature of the market. (The promotion of sports personalities and musicians as 'role models' for young black people seems particularly ill-judged in this respect.) Instead of wasting the crucial opening years of their working lives in this way, they would be much better off striving for careers in more modest but realistic occupations, such as teaching or engineering. Society in general would also benefit if larger numbers of people aimed to become teachers or engineers, whereas it will never suffer from a shortage of lawyers.

The idea that markets can be curbed seems more and more like a fantasy these days, but Frank and Cook propose a solution based on the psychology of status. Classical economics, based on the premise that individuals act rationally, assumes that wealth makes people happy according to its value. By this reckoning, the pleasure a man obtains from owning a £50,000 sports car resides principally in the absolute value of £50,000, which would be equally pleasing if it was realized in the form of a country cottage or a savings account. Frank and Cook argue that the man's satisfaction actually derives from seeing that his neighbours' cars cost less. What counts is not his car's absolute value, but its value relative to those of his near competitors.

One way to reduce the appeal of winner-take-all markets, say Frank and Cook, is to lower the attraction of large salaries by levying direct taxes that could reach marginal rates as high as 90 per cent. They have not reinvented progressive income tax, though. Their proposal is that people should be taxed on what they spend. This would be a visible hand that encouraged the rich to distribute wealth by investing it, perhaps creating jobs or other social goods thereby. They would be just as happy, because their cars would still be more expensive than those of their neighbours, although they might buy Porsches instead of costlier Ferraris.

There is nothing distinctively evolutionary about Frank and Cook's analysis – though Frank draws an analogy between winner-take-all markets and mating systems in which a single male monopolizes the females. Traditional psychology has recognized the importance of relative status without the help of Darwinian thinking. It resonates with evolutionary themes, however, and it extends lines of thought from Frank's earlier work, in which evolutionary themes were clearly visible. It therefore helps to integrate Darwinian theory with human affairs in an organic way, in which evolutionary ideas add depth, instead of erasing detail.

Robert Frank and Philip Cook have put their fingers on what wealth does for the rich. Richard Wilkinson's work reveals what the wealth of the rich does to those less well off. While the rich man's car makes him happy because it costs more than his neighbour's, it makes his neighbour not only less happy, but less healthy.

Unfortunately, the effect extends from neighbourhood to neighbourhood, growing in strength as it descends the social scale. Although the cost of the rich man's car will not shorten his neighbour's life noticeably, since his neighbour will be at least comfortably off, the effects of inequality upon the humblest will be grave. The greater the gap between the richest and the poorest members of society, the worse the effects will be. Unequal societies are unhealthy societies.

Unequal societies also appear to be something of a curiosity, evolutionarily speaking. In terms of hierarchy, agricultural and industrial societies resemble those of chimpanzees more than those of human foragers. Existing hunter-gatherers live in societies without classes, and without clear leadership. When humans developed farming, they seem to have got back in touch with their deep primate roots, recreating hierarchies and elaborating them into social classes.

Being aware that hunter-gatherers tend to kill each other frequently, anthropologists are not inclined to assume that co-operation in traditional societies is always benign and voluntary. They tend to construe nomadic egalitarianism as the result of what David Erdal and Andrew Whiten call counter-domination, and what the Australians call cutting down tall poppies. Although members of egalitarian hunter-gatherer societies will acknowledge those who perform well, and defer to them, they will resist any attempt to translate performance into dominance. In this respect, suggests Bruce Knauft, the critical difference between apes and people is that humans hate to grovel. Apes, by contrast, have a ready repertoire of submissive gestures with which to demonstrate their subordination.[45]

Christopher Boehm sees egalitarian traditional societies as ones in which the old primate forces are still at work, constantly inciting individuals to strive for a position above their fellows, but are held in check by stringent counter-measures. One chimpanzee may sometimes intervene to prevent a second from exerting too much dominance over a third, and subordinate males can hold considerable power within a group. In egalitarian human societies, says Boehm, individuals that attempt to reach too high above their fellows are subjected to a much more coherent form of pressure than the *ad hoc* measures which chimpanzees can muster. If one poppy grows taller than the rest, it will assuredly be lopped to size.

Human groups can achieve these coalitions because they can build a 'moral community', asserting the value of equality and implementing sanctions against those who transgress this value. Boehm suggests that the first moral community arose in response to the social problem of an excessively dominant alpha male, and that 'when the rank-and-file unanimously branded their too-dominating top male as a social undesirable, group-morality . . . was born'.[46]

Once established, moral communities could deal with other social problems, acting to suppress any form of deviance from group norms. Mechanisms such as these tend to treat innovation as deviance. Looking at a traditional society, a modern conservative would see eloquent proof that you don't get anywhere unless you let some of your children grow taller than others if they have it in them to do so.

If taller children take us further, then logic would seem to dictate that the taller they grow, the better. Whereas conservatism traditionally accepted inequality as part of the proper order of things, the modern Right is inclined to celebrate inequality as a sign that everybody is making the best of themselves. An unequal society is an enterprising one, and enterprise is healthy.

Even in societies that have done well from enterprise, however, people tend to remain uneasy about inequality. Outside the

United States, public opinion in most industrialized countries would probably have more sympathy with the hunter-gatherers than the neo-conservatives. But sympathies have little force against a vigorous ideology of self-interest which can back its arguments by pointing to a track record of success. To turn their sentiments into arguments, modern egalitarians need counter-data. These already exist, and Richard Wilkinson has compiled them into a counter-database.[47]

One of the most striking collections of evidence is in what have come to be known as the 'Whitehall studies', of 17,000 civil servants working in the British government's heartland. Their death rates adhered to the Civil Service grade structure as strictly as their pay scales. Overall mortality at the bottom of the hierarchy was three times higher than at the top. This could not be an effect of poverty, since although junior civil servants are not wealthy by contemporary standards, they are paid enough to support a life of modest comfort. Nor did it arise because the lower grades held to an unenlightened lifestyle, because smoking and cholesterol accounted for little of the gradient in fatal heart disease, which was four times more common at the bottom than the top.

Whitehall seems to be the world in microcosm. More than twenty studies have upheld the relationship between death rates and inequality; only two, sharing the same data, have not. Once a society has grown sufficiently prosperous that degenerative diseases replace infectious ones as its main causes of death, Richard Wilkinson observes, further increases in prosperity do not appear to affect health very much. Among the developed countries, the healthiest are not the richest societies, but the most egalitarian ones. They also have fewer murders and road accidents.[48]

From data about death and money, the network of researchers to which Wilkinson belongs has gone on to construct a powerful body of argument which identifies social cohesion as the basis of a healthy society. One of its showpieces was the small Pennsylvania

town of Roseto, whose very name indicates the source of its inhab-
itants' longstanding closeness. The town was built in the 1880s by
immigrants from the Italian town of Roseto, who maintained
notably strong family and neighbourhood ties in their new home.
By the 1930s, the incidence of death from heart disease was also
markedly lower than in neighbouring towns, despite the Rosetans'
taste for food cooked in lard. Individuals lost the benefit if they
moved away, suggesting that the effect was not genetic, and the
entire town lost it as the younger generation abandoned community values in favour of Cadillacs. They left behind a world in
which, as researchers who had known it in the early 1960s recalled,
many were affluent but nobody tried to keep up with the Joneses.

Roseto is the kind of small town that occupies a special place in
the American imagination, embodying an ideal believed to have
been lost to the ambient suburbs on the one hand and the hair-
trigger cities on the other. It seems to have more in common with
a Samoan village than a modern settlement. But such benefits of
social cohesion can also be discerned in Japan, long a by-word for
crowding and pressure of work. By the end of the 1980s, Japan
had both the highest life expectancy and the lowest income differences of any developed country. The former had risen,
spectacularly, since the 1960s, while the latter had fallen. We do
not need to reinvent a life with ancestral undertones, whether
modelled on small-town USA, Samoan villages, or even hunter-
gatherer bands, in order to enjoy social harmony and the good
health that comes with it.

The data that Wilkinson presents are like the evidence that
people's health can be damaged by passive smoking. They transform the terms of the debate about freedom and responsibility.
Smokers and the industry that supplies them made some head-
way by arguing that they were harming only themselves, and
measures to restrict their behaviour infringed their personal liberty. The medical evidence dispelled the myth that smokers
smoke in a social vacuum, and likewise Wilkinson's evidence

dispels the myth that personal wealth is a matter for the individual alone. According to the traditional view from below, the rich obtained their wealth at the expense of the poor. This conviction has lost much of its force in societies where affluence is general and even the poorest households are likely to have consumer goods considered luxuries in earlier generations. The health evidence against inequality reasserts that the rich enjoy their wealth at the expense of the poor, and that the price paid by the worse off increases with every cut in high earners' income tax and every winner-take-all deal. Those in the middle of the heap, who may have thought of themselves as beneficiaries of modern economic arrangements, should note the lesson of the Whitehall studies: the effect extends all the way up the scale.

As a political argument, the most remarkable thing about Wilkinson's vision is that although it starts from conventional data about health and wealth, its focus is not material but psychological. By drawing upon data from different continents, it implicitly acknowledges that the psychological processes it describes may be universal, and therefore part of an evolved human nature. They are certainly only too familiar. The power of status is felt in the peculiarly shrill ache of shame or disrespect. A recollected embarrassment or humiliation can sting almost as sharply as when it happened, but the recollection of a famous social triumph is usually just a memory. Part of our minds seems incapable of treating any slight as petty. The consequences are clear when the aggrieved party is an urban youth with a gun and little else going for him. The small change of status may also be lethal at less abject levels in the social hierarchy, including ones which appear quite comfortable. It just takes longer.

At the gut level, our distant cousins appear to feel as we do. Wilkinson has drawn on primate studies, particularly those conducted by Robert Sapolsky on baboons. Subordinate baboons have elevated levels of cortisol, which suggests that primates in general may respond in a similar physiological way to the oppressive effects of hierarchy. The effect is to make them grow old more quickly. Robert Wright cited low cortisol levels among rural Samoans to support his claim that 'going crazy', in the form of becoming depressed, isn't natural. It would appear to be natural among baboons, though, which suggests that if our ancestors really were free from depression, they had developed strategies we would do well to identify.

We might not want to emulate them, though. Equality does not appear to have the ameliorative effect on hunter-gatherer homicide rates that it does in modern society; and Christopher Boehm's moral communities are not conducive either to personal liberty or to the diversity of thought needed to meet the challenges of a rapidly changing world. But with depression bidding to top the world illness league, we need to know as much about its history as we can.

The same is true for inequality and social cohesion. Wilkinson's work is imbued with the same spirit as explicitly evolutionary theory. It trades in the same psychological currency as modern Darwinism, and it recognizes that we are primates. The missing link between the two currents of thought is an account of how equality rose and fell in the hominid lineage, after the split with our nearest relatives. This might seem like a merely academic exercise, were it not for the lesson of the inequality data; that a good society is an equal one.

A *firm grip* is needed.

You don't get much out of a set of tools if you are scared of using them. Yes, you can do damage with a saw and a drill, but you are likely to do more damage and achieve less if you do not apply them with resolve. If you deal with the truth of an evolved mind by constantly querying and qualifying any possible evolutionary insights, seeking not to strengthen but to dilute them, you are never really likely to be satisfied with the results. Better to follow Darwin's example, and dare to ask.

Darwin's courage had its limits, it is true. He delayed fifteen years before publishing *The Origin of Species*. But if a Victorian naturalist of retiring disposition could mount a challenge to orthodoxy more profound than the notion that the Earth moves around the Sun, modern lay people in secularized societies should not be afraid to follow it through.

3

When somebody tells you that it is better to grasp unpleasant truths than to remain ignorant, the propositions they usually have in mind are ones that you may find unpleasant, but they don't. There are plenty of evolutionary commentators who will shake crocodile heads as they talk of men's xenophobic disposition, or women's reluctance to compete for status, or the strains that monogamy imposes upon husbands. There is plenty in evolutionary theory to warm the cockles of the conservative heart, which believes it is noble for men to fight under flags, and proper that women should stay at least half in the home, and that male philandering should be looked upon with an indulgent eye.

In a period when the watchword is reduced expectations, it may be tempting for observers on the Left or in the centre ground to sign up for this package, though their degree of reluctance may vary. Left-wing arguments used to rise blithely into the clouds, but nowadays all but the immutable fringe seem to descend inexorably into bathos. Even Peter Singer, whose book *Animal*

Liberation was the founding text of the animal rights movement, goes down this road in his pamphlet *A Darwinian Left?*[249] His concluding outline of such a movement describes a Left which would recognize that human nature is neither inherently good nor infinitely malleable. It would accept that some inequalities are not caused by 'discrimination, prejudice, oppression or social conditioning' – including those in the boardroom, which, he would agree, may be due to an innately greater readiness on the part of men to pursue career goals. A Darwinian Left would expect people to compete in pursuit of their own interests, or those of their kin, but also to respond positively to offers of co-operation, when satisfied that these are genuine. Like the traditional Left, it would stand up for the weak, the poor and oppressed, 'but think very carefully about what social and economic changes will really work to benefit them'.

Singer's final remarks, in which he suggests a distant prospect of transcending human nature through genetic engineering, read like a gesture to the left-wing tradition of bringing an essay to rest in the promised land. His programme may be eminently realistic, but its few distinctively left-wing elements are so diffident that they could be incorporated quite comfortably into the projects of most conservative parties. 'In some ways,' he admits, 'this is a sharply deflated vision of the left, its utopian ideas replaced by a coolly realistic view of what can be achieved. That is, I think, the best we can do today.'

It is a little dismaying to hear a philosopher-activist, who works both to interpret the world and to change it, sounding like a politician talking down manifesto commitments. Accepting the existence of an evolved human nature does not mean accepting conservative readings of how adaptations play out in modern life. In the case of the glass ceiling, a richer evolutionary perspective would recognize the possibility that male coalitions try to exclude those women who are prepared to pursue their careers as single-mindedly as men. In most other cases, evolutionary thought generates ample material for both the Left and the Right, and diverse other persuasions.

This is not to suggest that one evolutionary argument is as good as another, or that we should pick and choose among them according to our ideological inclinations. Evolutionary theory is a rich and diverse environment. Casual visitors can come away with snapshots of what they want to see, but that is just tourism. It takes an open mind to understand an evolved mind.

Making use of evolutionary thought does not, however, require you to buy a bundle of evolutionary theory, put it on your desk and stare at it. Hardly anybody would go into a shop and ask for an electric motor, but people buy them in huge numbers without even realizing. Electric motors are incorporated into other machines, from cars and refrigerators to computers and hairdriers. Microprocessors have already infiltrated themselves into devices such as telephones, cameras and washing machines. It is probably only a matter of time before they embed themselves in hairdriers. Evolutionary ideas may go the same way, becoming elements of social and psychological thought so commonplace that they are no longer advertised.

Irreconcilable evophobes will shudder at this prospect, preferring to have modern Darwinism out in the open where they can cast aspersions on it. They will worry, rightly, that hypotheses will be used as axioms, and that dubious arguments will merge into the background of received wisdom. One rejoinder to this kind of concern would be to warn that the science will develop anyway, as will its public profile, so liberals should become involved in order to wrestle the reactionaries and determinists for influence. If that's what it takes, fine, but this belongs at the bottom of the list of reasons to join the project of developing an evolutionary psychology.

Above it is a series of reasons to feel confident about human possibilities. In the next slot up goes the suggestion that even if you feel obliged to see innate knowledge as a network of constraints, human freedom shows up all the brighter against a dull background of limits. The exuberance of human psychological and social diversity affirms that whatever our limits actually are, our freedom is breathtaking.

Beyond this level come the new possibilities that can be discerned from current trends in thought about the evolved mind. One is that streams of knowledge that have been flowing further apart will now come closer together. Edward O. Wilson has hailed this trend and called it 'consilience'.[50] His thesis develops the theme that he unveiled in 1975, to outrage in some quarters, as *Sociobiology: The New Synthesis*. Wilson foresees that the social sciences will split, with one part merging into the humanities and another 'folding into or becoming continuous with biology'. At the same time, the humanities themselves will draw closer to science.

Developments along these lines could be compatible with the kind of equal partnerships between disciplines on which this book is premised. With a place at the centre of these partnerships, evolutionary science could claim to be first among equals. But 'folding into' suggests an acquisition rather than a joint venture, and Wilson seems to see the role of science in general as one of leadership. He goes so far as to propose that it can replace religion in meeting humankind's need for the sacred, although the signs so far are unpropitious. Science can provoke people's sense of wonder, but it does not inspire the sense of awe that people reserve for what they feel to be truly sacred.

Among the specialists, likewise, evolutionary thinking will be held in increasingly high regard. Even the converts who anticipate a Darwinian revolution are not going to dump the tools with which they served their apprenticeships. Evolutionary appliances will help unify knowledge about artefacts and knowledge about fossils, but archaeologists are not going to throw away their trowels and run computer simulations instead. They know there is no substitute for getting their boots dirty. Nor will social scientists abandon the methods they have developed for studying society, especially since evolutionary scientists will be telling them that society is what caused the mind to evolve in the first place.

And it is in society at large, not in the private self or knowledge in the abstract, where the greatest benefits stand to be earned from an evolutionary approach to the mind. As Peter Singer affirms, it is necessary to understand human nature in order to bring out the best in it. Evolutionary psychology and allied lines of inquiry have already identified as key themes fairness, co-operation, differences of interests between the sexes, and equality. Those who want a fairer, more co-operative and less unequal society should gain confidence about what is possible as they become used to handling the tools that sociobiological studies make available.

Evolutionary examples relevant to these key themes tend to depict imperfect settlements between closely balanced forces. Bonobo males remain typically male, but their style is cramped by female solidarity. Hunter-gatherer males retain the urge to rise above their fellows, but each time one tries it, the rest cut him down to size. The kind of social reforms that evolutionary thinking may suggest will have a rather different tenor, being obliged to observe modern civil proprieties, but they will be imperfect compromises too. That will be a necessary condition of their success, and may in itself be reassuring after a century of discredited utopias. Coming to terms with an evolved mind should help us work out ways in which the world could be made a better place, but it won't be the end of the world as we know it.

Notes and References

Note: Because the Internet will be more accessible to many readers than academic libraries, World Wide Web addresses are included where known. The Web changes fast, however, so the references are no guarantee that the resources will still be available when requested.

One   Costs

1 Daniel C. Dennett (1995), *Darwin's Dangerous Idea: Evolution and the meanings of life*, Allen Lane, London.
2 Associated Press, 27 May 1998.
3 Michael White and John Gribbin (1998), *Stephen Hawking: A life in science*, Penguin, London.
4 Steven Rose, Richard Lewontin and Leon Kamin (1990), *Not In Our Genes: Biology, ideology and human nature*, Penguin, London.
5 Donald Symons (1992), 'On the use and misuse of Darwinism in the study of human behavior', in Jerome H. Barkow, Leda Cosmides and John Tooby (eds) *The Adapted Mind: Evolutionary psychology and the generation of culture*, Oxford University Press, New York.
6 Cf. John Horgan (1995), 'The new Social Darwinists', *Scientific American* 273(4), 150–7.
7 Devendra Singh, n.d, lecture abstract; (1993), 'Adaptive significance of female physical attractiveness: role of waist-to-hip ratio', *Journal of Personality and Social Psychology* 65(2), 293–307; (1994), 'Waist-to-hip ratio and judgment of attractiveness and healthiness

of female figures by male and female physicians', *International Journal of Obesity* 18, 731–7.

8 Horgan, op. cit.

9 Olga Soffer (1987), 'Upper Paleolithic connubia, refugia, and the archaeological record from Eastern Europe', in O. Soffer (ed.), *The Pleistocene Old World: Regional perspectives*, Plenum, New York; Olga Soffer and N. D. Praslov (1993), *From Kostenki to Clovis: Upper Paleolithic–Paleo-Indian adaptations*, Plenum, New York; Margaret W. Conkey (1997), 'Mobilizing Ideologies: Paleolithic "art", gender trouble, and thinking about alternatives', in Lori D. Hager *Women in Human Evolution*, Routledge, London.

10 Fielding H. Garrison (1929), *An Introduction to the History of Medicine*, W. B. Saunders, Philadelphia.

11 LeRoy McDermott (1996), 'Self-representation in Upper Paleolithic female figurines', *Current Anthropology* 73, 227–75.

12 'NYU anthropologist says female statuettes from Ice-Age Europe were carved to protect health of pregnant mothers' (press release), Eurekalert, American Association for the Advancement of Science
http://www.eurekalert.org/releases/nyu-asfsf.html

13 Douglas W. Yu and Glenn H. Shepard Jr (1998), 'Is beauty in the eye of the beholder?' *Nature* 396, 321–2.

14 Jerome H. Barkow, Leda Cosmides and John Tooby (1992), op. cit.; Leda Cosmides and John Tooby (1997), 'Evolutionary Psychology: A primer', Center for Evolutionary Psychology, University of California Santa Barbara
http://www.psych.ucsb.edu/research/cep/primer.htm

15 Martin Daly and Margo Wilson (1988), *Homicide*, Aldine de Gruyter, New York.

16 Martin Daly and Margo Wilson (1996), 'Violence against stepchildren', *Current Directions in Psychological Science* 5(3),77–81.

17 *New York Times* 7 August 1993.

18 Martin Daly and Margo Wilson (1998), *The Truth About Cinderella: A Darwinian view of parental love*, Weidenfeld & Nicolson, London.

19 Marek Kohn, 'Cinderella revisited', *Independent on Sunday*, London, 24 November 1996.

20 Ronald Walters (n.d.), 'The Politics of the Federal Violence Initiative', Howard University (unpublished draft); Pat Shipman (1994), *The Evolution of Racism*, Simon & Schuster, New York.

21 Paul Bahn (1996), 'New developments in Pleistocene art', *Evolutionary Anthropology* 4(6), 204–15.

22 Stephen Jay Gould, 'Evolution: The pleasures of pluralism', *New York Review of Books* 26 June 1997.

23 Rudyard Kipling (1902), 'How the Leopard Got His Spots', in *World's Greatest Classic Books*, Corel, 1995, Ottawa (CD-ROM).

24 Amotz and Avishag Zahavi (1997), *The Handicap Principle: A missing piece of Darwin's puzzle*, Oxford University Press, Oxford.

25 John Brockman (1995), *The Third Culture: Beyond the scientific revolution*, Simon & Schuster, New York.

26 Brockman, op. cit.; Stephen Jay Gould, 'Darwinian Fundamentalism', *New York Review of Books* 12 June 1997.

27 David Sloan Wilson and Elliott Sober (1994), 'Reintroducing group selection to the human behavioral sciences', *Behavioral and Brain Sciences* 17, 585–654; John Maynard Smith (1998), 'The origin of altruism', *Nature* 393, 639–40.

28 Lynn Hunt, 'Send in the clouds', *New Scientist* 30 May 1998; W. D. Hamilton and T. M. Lenton (1998), 'Spora and Gaia: how microbes fly with their clouds', *Ethology, Ecology and Evolution* 10, 1–16; http://www.unifi.it/unifi/dbag/eee/; Robert J. Charlson et al. (1987), 'Occanic phytoplankton, atmospheric sulphur, cloud albedo and climate', *Nature* 326, 655–61.

29 Kenan Malik, 'The Beagle sails back into fashion', *New Statesman* 6 December 1996.

30 'Matters of life and death: the world view from evolutionary psychology', *Demos Quarterly* issue 10, 1996.

31 Ivan Turk, Janez Dirjec and Boris Kavur, 'The oldest musical instrument in Europe discovered in Slovenia?' http://www.zrc-sazu.si/www/iza/piscal.html

32 University of Witwatersrand press release, 'Major fossil find at Sterkfontein caves', 9 December 1998, http://www.wits.ac.za/press_releases/clarke.html

33 M. J. Morwood et al. (1998), 'Fission-track ages of stone tools and fossils on the east Indonesian island of Flores'. *Nature* 392, 173.

34 C. B. Stringer (1994), in Steve Jones, Robert Martin and David Pilbeam (eds) *Cambridge Encyclopedia of Human Evolution*, Cambridge University Press, Cambridge.

35 C. C. Swisher III et al. (1996), 'Latest *Homo erectus* of Java:

Potential contemporaneity with *Homo sapiens* in Southeast Asia', *Science* 274, 1870–4.

36 Tomas Lindahl (1997), 'Facts and artifacts of ancient DNA', *Cell* 90, 1–3; Matthias Krings et al., 'Neanderthal DNA sequences and the origin of modern humans', *Cell* 90, 19–30.

Two   Symmetry

1 John Wymer (1968), *Lower Palaeolithic Archaeology in Britain as represented by the Thames Valley*, Humanities Press, New York / John Baker, London.

2 The English Rivers Palaeolithic Survey Project, http://www.eng-h.gov.uk/ArchRev/rev94_5/engrivs.htm

3 William Watson (1968), *Flint Implements: An account of stone age techniques and cultures*, British Museum, London; Walter Shepherd (1972), *Flint: its origins, properties and uses*, Faber & Faber, London.

4 B. Asfaw et al. (1992), 'The earliest Acheulean from Konso-Gardula', *Nature* 360, 732–5; Eric Delson (1997), 'One skull does not a species make', *Nature* 389, 445–6.

5 Glynn Isaac (1986), 'Foundation Stone: early artefacts as indicators of activities and abilities', in G. N. Bailey and P. Callow (eds), *Stone Age Prehistory: Studies in memory of Charles McBurney*, Cambridge University Press, Cambridge.

6 J. A. J. Gowlett (1986), 'Culture and conceptualisation: the Oldowan–Acheulian gradient', in Bailey and Callow, op. cit.

7 S. Semaw et al. (1997), '2.5-million-year-old stone tools from Gona, Ethiopia', *Nature* 385, 333–6; W. H. Kimbel et al., 1996, 'Late Pliocene *Homo* and Oldowan tools from the Hadar Formation (Kada Hadar Member), Ethiopia', *Journal of Human Evolution* 31, 549–61.

8 Kathy D. Schick and Nicholas Toth (1993), *Making Silent Stones Speak: Human evolution and the dawn of technology*, Weidenfeld & Nicolson, London.

9 Glynn Isaac (1986), op. cit.

10 Eileen O'Brien (1981), 'The projectile capabilities of an Acheulian handaxe from Olorgesailie', *Current Anthropology* 22(1), 76–9; Glynn Isaac (1977), *Olorgesailie*, University of Chicago Press, Chicago.

11 William H. Calvin (1993), 'The unitary hypothesis: a common neural circuitry for novel manipulations, language, plan-ahead, and throwing?' in Kathleen R. Gibson and Tim Ingold (eds), *Tools,*

Language and Cognition in Human Evolution, Cambridge University Press, Cambridge; Calvin (1990), *The Ascent of Mind: Ice Age climates and the evolution of intelligence*, Bantam, New York; http://weber.u.washington.edu/~wcalvin/bk5/bk5ch8.htm

12 Iain Davidson and William Noble (1993), 'Tools and language in human evolution', in Gibson and Ingold, op. cit.

13 Bruce Bradley and C. Garth Sampson (1986), 'Analysis by replication of two Acheulean artefact assemblages', in Bailey and Callow, op. cit.

14 Brian Hayden (1993), 'The cultural capacities of Neanderthals: a review and re-evaluation', *Journal of Human Evolution* 24, 113–46.

15 F. F. Wenban-Smith (1989), 'The use of canonical variates for determination of biface manufacturing technology at Boxgrove Lower Palaeolithic site and the behavioural implications of this technology', *Journal of Archaeological Science* 16,17–26.

16 Ibid.

17 Clive Gamble (1995), 'Personality most ancient', *British Archaeology* no. 1, 6; http://britac3.britac.ac.uk/cba/ba/ba19/ba1feat.html

18 Mark Roberts (1996), 'And then came clothing and speech', *British Archaeology* no. 19, 8–9; http://britac3.britac.ac.uk/cba/ba/ba19/ba19feat.html

19 Jared Diamond (1991), *The Rise and Fall of the Third Chimpanzee: How our animal heritage affects the way we live*, Radius, London.

20 Neville Agnew and Martha Demas (1998), 'Preserving the Laetoli footprints', *Scientific American* 279(3), 26–37.

21 Henry M. McHenry and Lee R. Berger (1998), 'Body proportions in *Australopithecus afarensis* and *A. africanus* and the origin of the genus *Homo*', *Journal of Human Evolution* 35, 1–22; Lee Berger (1998), 'Redrawing our family tree?' *National Geographic* 194(2), 90–9.

22 F. Schrenk et al. (1993), 'Oldest *Homo* and Pliocene biogeography of the Malawi Rift', *Nature* 365, 833–6; Andrew Hill et al. (1992), 'Earliest *Homo*', *Nature* 355, 719–22.

23 Bernard Wood (1992), 'Origin and evolution of the genus *Homo*', *Nature* 355, 783–90.

24 Richard Leakey and Roger Lewin (1992), *Origins Reconsidered*, Little, Brown, London.

25 Interview with author, 5 November 1992.

26 Roy Larick and Russell L. Ciochon (1996), 'The African emergence

and early Asian dispersals of the genus *Homo*', *American Scientist* 84(6), 538–51.

27 Concepción Borja et al. (1997), 'Immunospecificity of albumin detected in 1.6 million-year-old fossils from Venta Micena in Orce, Granada, Spain', *American Journal of Physical Anthropology* 103, 433–41.

28 J. M. Bermúdez de Castro et al. (1997), 'A hominid from the Lower Pleistocene of Atapuerca, Spain: possible ancestor to Neanderthals and modern humans', *Science* 276, 1392–5; 'A new face for human ancestors', same issue, 1331–3.

29 Julia A. Lee-Thorp, Nikolaas J. van der Merwe, C. K. Brain (1994), 'Diet of *Australopithecus robustus* at Swartkrans from stable carbon isotopic analysis', *Journal of Human Evolution* 27, 361–72.

30 Robert Foley (1995), *Humans Before Humanity*, Blackwell, Oxford.

31 Henry M. McHenry (1982), 'The pattern of human evolution: studies on bipedalism, mastication, and encephalization', *Annual Review of Anthropology* 11: 151–73;
http://www-anthro.ucdavis.edu/faculty/mchenry/bipedal.htm

32 Ibid.

33 Robert Foley and Phyllis C. Lee (1996), 'Finite social space and the evolution of human social behaviour', in James Steele and Stephen Shennan (eds), *The Archaeology of Human Ancestry: Power, Sex and Tradition*, Routledge, London; (1989), 'Finite social space, evolutionary pathways and reconstructing hominid behavior', *Science* 243, 901–6.

34 Birgitta Sillén-Tullberg and Anders P. Møller (1993), 'The relationship between concealed ovulation and mating systems in anthropoid primates: a phylogenetic analysis', *American Naturalist* 141, 1–25.

35 Henry M. McHenry (1996), 'Sexual dimorphism in fossil hominids and its socioecological implications', in Steele and Shennan, op. cit.

36 J. Michael Plavcan and Carel P. van Schaik (1997), 'Interpreting hominid behavior on the basis of sexual dimorphism', *Journal of Human Evolution* 32, 345–74.

37 Bernard Wood (1997), 'The oldest whodunnit in the world', *Nature* 385, 292–3; S. Semaw et al., '2.5-million-year-old stone tools, from Gona, Ethiopia', ibid. 333–6.

38 Kathleen R. Gibson (1993), 'Tool use, language and social behavior in relationship to information processing capacities', Gibson and

Ingold, op. cit.; Nicholas Toth et al. (1993), 'Pan the Tool-maker: investigations into the stone tool-making and tool-using capabilities of a bonobo (*Pan paniscus*)', *Journal of Archaeological Science* 20, 81–91.

39 Mzalendo Kimunjia (1994), 'Pliocene archaeological occurrences in the Lake Turkana basin', *Journal of Human Evolution* 27, 159–71.

40 Roy Larick and Russell L. Ciochon, op.cit.

41 M. J. Rogers, C. S. Feibel and J. W. K. Harris (1994), 'Changing patterns of land use by Plio-Pleistocene hominids in the Lake Turkana basin', *Journal of Human Evolution* 27, 139–59.

42 Michael Pitts and Mark Roberts (1997), *Fairweather Eden*, Century, London.

43 P. R. Jones (1995), 'Results of experimental work in relation to the stone industries of Olduvai Gorge', in M. Leakey and D. Roe (eds), *Olduvai Gorge* vol. 5: *Excavation in Beds III–IV and the Masele Beds 1968–71*, Cambridge University Press, Cambridge.

44 J. C. Mitchell (1995), 'Studying biface utilisation at Boxgrove: roe deer butchery with replica handaxes', *Lithics* 16, 64–9.

45 Peter R. Jones (1980), 'Experimental butchery with modern stone tools and its relevance for Palaeolithic archaeology', *World Archaeology* 12(2), 153–65.

46 Toshisada Nishida et al. (1992), 'Meat-sharing as a coalition strategy by an alpha-male chimpanzee', in T. Nishida et al. (eds), *Topics in Primatology*, vol. 1, *Human Origins*, University of Tokyo Press, Tokyo.

47 Craig B. Stanford et al. (1994), 'Hunting decisions in wild chimpanzees', *Behaviour* 131, 1–18.

48 H. Kaplan and K. Hill (1985), 'Hunting ability and reproductive success among male Ache foragers', *Current Anthropology* 26, 131–3.

49 John D. Speth (1994), 'Early hominid hunting and scavenging: the role of meat as an energy source', *Journal of Human Evolution* 18, 329–43.

50 Robert A. Foley and Phyllis C. Lee (1991), 'Ecology and energetics of encephalization in hominid evolution', *Philosophical Transactions of the Royal Society*, London Series B, 334, 223–32.

51 Alexander Cockburn (1996), 'A short, meat-oriented history of the world. From Eden to the Mattole', *New Left Review* 215, 16–42.

52 Leslie C. Aiello and Peter Wheeler (1995), 'The expensive-tissue hypothesis: the brain and the digestive system in human and

primate evolution', *Current Anthropology* 36(2), 199–221.

53 Dean Falk (1995), *Current Anthropology* 36(2), 212; Steven R. Leigh (1992), 'Cranial capacity evolution in *Homo erectus* and early *Homo sapiens*', *American Journal of Physical Anthropology* 87, 1–13.

54 S. T. Parker and K. R. Gibson (1979), 'A developmental model for the evolution of language and intelligence in early hominids', *Behavioral and Brain Sciences* 2, 367–408.

55 T. H. Clutton-Brock and P. H. Harvey (1980), 'Primates, brains and ecology', *Journal of Zoology* 190, 309–23.

56 Alison Jolly (1966), 'Lemur social behavior and primate intelligence', *Science* 153, 501–6.

57 N. K. Humphrey (1976), 'The social function of intellect', in P. P. G. Bateson and R. A. Hinde (eds), *Growing Points in Ethology*, Cambridge University Press, Cambridge.

58 André Leroi-Gourhan, '140 thousand years of continuous human occupation on a single site', *Illustrated London News* 29 November 1952.

59 Lewis R. Binford (1981), *Bones: Ancient men and modern myths*, Academic Press, New York.

60 Hartmut Thieme (1997), 'Lower Palaeolithic hunting spears from Germany', *Nature* 385, 807–10; also 'News & Views', ibid., 767–8.

61 Amotz and Avishag Zahavi (1997), *The Handicap Principle: A missing piece of Darwin's puzzle*, Oxford University Press, Oxford.

62 Alan Grafen (1990), 'Biological signals as handicaps', *Journal of Theoretical Biology* 144, 517–46.

63 Carl T. Bergstrom, 'The theory of costly signalling',
    http://calvino.biology.emory.edu/signalling/week4contents.html

64 Thorstein Veblen, *Theory of the Leisure Class*, American Studies @ the University of Virginia,
    http://xroads.virginia.edu/~HYPER/VEBLEN/veb_toc.html

65 Geoffrey F. Miller (1997), 'How mate choice shaped human nature: A review of sexual selection and human evolution', in C. Crawford and D. L. Krebs (eds), *Handbook of Evolutionary Psychology: Ideas, issues and applications*, pp. 87–129, Lawrence Erlbaum, Mahwah, New Jersey;
    http://ada.econ.ucl.ac.uk/papers/sex.htm

66 D. C. Johanson and M. A. Edey (1981), *Lucy: The beginnings of humankind*, Granada, St Albans. Quoted in *Blood Relations*.

67 M. S. Dawkins and T. Guilford (1991), 'The corruption of honest signalling', *Animal Behaviour* 41, 865–73.

68 Steve Jones (1993), *The Language of the Genes*, HarperCollins, London.

69 Jane Goodall (1986), *The Chimpanzees of Gombe: Patterns of behavior*, Belknap Press, Harvard University Press, Cambridge, Mass.

70 J. Michael Plavcan and Carel P. van Schaik (1997), op. cit.

71 Camilla Power and Leslie Aiello (1997), 'Female proto-symbolic strategies', in Lori D. Hager (ed.) (1997), *Women in Human Evolution*, Routledge, London.

72 C. O. Lovejoy (1981), 'The origin of man', *Science* 211, 341–50.

73 In Lori D. Hager (ed.), op. cit.

74 Pascal Gagneux, David S. Woodruff and Christophe Boesch (1997), 'Furtive mating in female chimpanzees', *Nature*, 387, 358–9.

75 Charles Darwin, 'The Descent of Man', University of Sheffield, http://www.shef.ac.uk/uni/projects/gpp/Tapestry/science/thedes1.html

76 Robert A. Foley and Phyllis C. Lee (1991), op. cit.

77 Christophe Boesch (1991), 'Teaching among wild chimpanzees', *Animal Behaviour* 41, 530–2; (1993), 'Aspects of transmission of tool use in wild chimpanzees', Gibson and Ingold, op. cit.

78 Geoffrey F. Miller, op. cit.

79 M. B. Roberts, S. A. Parfitt, M. I. Pope, F. F. Wenban-Smith (1997), 'Boxgrove, West Sussex: rescue excavations of a Lower Palaeolithic landsurface (Boxgrove Project B, 1989–91)', *Proceedings of the Prehistoric Society* 63, 303–58.

80 Peter B. deMenocal (1995), 'Plio-Pleistocene African climate', *Science* 270, 53–9.

81 Richard Dawkins (1989), *The Selfish Gene*, Oxford University Press, Oxford.

82 Amotz and Avishag Zahavi, op. cit.

83 Armand Marie Leroi, 'Sing, prance, ruffle, bellow, bristle and ooze', *London Review of Books* 17 September 1998.

84 Malte Andersson (1994), *Sexual Selection*, Princeton University Press, Princeton, N.J.

85 Bruce Bradley and C. Garth Sampson (1986), op. cit.

86 Marek Kohn and Steven Mithen (1999), 'Handaxes: products of sexual selection', *Antiquity* 73, 518–26.

87 Steven Mithen (1994), 'Technology and society during the Middle Pleistocene: hominid group size, social learning and industrial

variability', *Cambridge Archaeological Journal* 4, 3–33; John McNabb and Nick Ashton (1995), 'Thoughtful flakers', *Cambridge Archaeological Journal* 5, 289–301.

88 Barbara J. King and Stuart G. Shanker (1997), 'The expulsion of primates from the garden of language', *Evolution of Communication* 1(1), 59–99.

89 Robert Foley (1996), 'The adaptive legacy of human evolution: a search for the environment of evolutionary adaptedness', *Evolutionary Anthropology* 4(6), 194–203.

90 Robert Foley (1995), op. cit; Robert A. Foley and Phyllis C. Lee (1991), op. cit.

91 Virginia Morrell (1995), *Ancestral Passions: The Leakey family and the quest for humankind's beginnings*, Simon & Schuster, New York.

92 Anna Barker (1998), 'A quantitative study of symmetry in hand-axes', unpublished MA dissertation, University of Reading.

93 P. R. Jones (1995), op. cit.

94 Thomas Wynn (1979), 'The intelligence of later Acheulean hominids', *Man* 14, 371–91.

95 Thomas Wynn (1981), 'The intelligence of Oldowan hominids', *Journal of Human Evolution* 10, 529–41; (1979), op. cit.

96 J. A. J. Gowlett (1996), 'Culture and conceptualisation: the Oldowan–Acheulian gradient', in Paul Mellars and Kathleen Gibson (eds), *Modelling the Early Human Mind*, McDonald Institute for Archaeological Research, Cambridge.

97 Annette Karmiloff-Smith (1992), *Beyond Modularity: A developmental perspective on cognitive science*, MIT Press, Cambridge, Mass.

98 Jerry A. Fodor (1983), *The Modularity of Mind: An essay on faculty psychology*, MIT, Cambridge, Mass.

99 Dan Sperber (1994), 'The modularity of thought and the epidemiology of representations', in L. A. Hirschfeld and S. A. Gelman (eds), *Mapping the Mind: Domain specificity in cognition and culture*, Cambridge University Press, Cambridge.

100 Ibid.; Jerry Fodor (1998), 'The trouble with psychological Darwinism', *London Review of Books* 20(2);
http://www.lrb.co.uk/v20n02/fodo2002.html

101 Steven Pinker and Paul Bloom (1990), 'Natural language and natural selection', *Behavioral and Brain Sciences* 13, 707–84; Stephen Jay Gould (1993), 'Tires and sandals', in *Eight Little Piggies*, Jonathan Cape, London.

102 Annette Karmiloff-Smith (1992), op. cit.

Three Trust

1 Alexander Marshack (1996), 'A Middle Paleolithic symbolic composition from the Golan Heights: the earliest known depictive image', *Current Anthropology* 37(2), 357–64; (1997), 'The Berekhat Ram figurine: a late Acheulian carving from the Middle East', *Antiquity* 71(272), 327–37,
   http://intarch.ac.uk/antiquity/marshack.html
2 André Leroi-Gourhan, '140 thousand years of continuous human occupation on a single site', *Illustrated London News* 29 November 1952.
3 Robert G. Bednarik (1995), 'Concept-mediated marking in the Lower Palaeolithic', *Current Anthropology* 36(4), 605–34, with comments and reply.
4 John Halverson (1992), 'The first pictures: perceptual foundations of Paleolithic art', *Perception* 21, 389–404.
5 E. S. Savage-Rumbaugh et al. (1996), 'Language perceived: *Paniscus* branches out', in W. C. McGrew, L. Marchant and T. Nishida (eds), *Great Ape Societies*, Cambridge University Press, Cambridge.
6 John Halverson, op. cit.
7 Alexander Marshack (1996), op. cit.
8 Alexander Marshack (1991), 'The Taï plaque and calendrical notation in the Upper Palaeolithic', *Cambridge Archaeological Journal* 1, 25–61.
9 David Lewis-Williams (1991), 'Wrestling With Analogy', *Proceedings of the Prehistoric Society* 57(I), 149–62.
10 Oliver Sacks (1993), *Migraine* (revised and expanded edition), Picador, London.
11 Jean Clottes and David Lewis-Williams (1996), 'Upper Palaeolithic cave art: French and South African collaboration', *Cambridge Archaeological Journal* 6(1) 137–9.
12 John Halverson, op. cit.
13 Francesco d'Errico et al. (1998), 'Neanderthal acculturation in Western Europe? A critical review of the evidence and its interpretation, with CA* comment', *Current Anthropology* 39, Supplement, Special Issue: *The Neanderthal problem and the evolution of human behavior*, S1–44.
14 Robert Foley and Marta Mirazón Lahr (1997), 'Mode 3

technologies and the evolution of modern humans', *Cambridge Archaeological Journal* 7(1), 3–36.

15 Alexander Marshack (1996), op. cit.; (1989), 'Evolution of the human capacity: the symbolic evidence', *Yearbook of Physical Anthropology*, 321–34.

16 Jared Diamond, op. cit.

17 William Golding (1955), *The Inheritors*, Faber & Faber, London.

18 Camilla Power and Ian Watts (1996), 'Female strategies and collective behaviour: the archaeology of earliest *Homo sapiens sapiens*', in Steele and Shennan, op. cit.; Chris Knight, Camilla Power and Ian Watts (1995), 'The human symbolic revolution: a Darwinian account', *Cambridge Archaeological Journal* 5(1), 75–114.

19 Alexander Marshack (1989), op. cit.

20 Chris Stringer and Robin McKie (1996), *African Exodus: The origins of modern humanity*, Jonathan Cape, London.

21 Richard F. Kay, Matt Cartmill and Michelle Balow (1998), 'The hypoglossal canal and the origin of human vocal behavior', *Proceedings of the National Academy of Sciences* 95(9), 5417–9; Kenneth Chang, 'When were the first words?' ABCNEWS.com, 27 April 1998,
http://www.abcnews.com/sections/science/dailynews/speech980 427.html

22 Merlin Donald (1991), *Origins of the Modern Mind: Three stages in the evolution of culture and cognition*, Harvard University Press, Cambridge, Mass.; Christopher Stringer and Clive Gamble (1993), *In Search of the Neanderthals: Solving the puzzle of human origins*, Thames & Hudson, London; Terrence Deacon (1997), *The Symbolic Species: The co-evolution of language and the human brain*, W. W. Norton, New York.

23 Merlin Donald (1994), 'Précis of "Origins of the modern mind"', *Behavioral and Brain Sciences* 16, 737–91; (1991), op. cit.

24 D. L. Cheney and R. M. Seyfarth (1990), *How monkeys see the world: Inside the mind of another species*, Chicago University Press, Chicago; Marc D. Hauser (1996), *The Evolution of Communication*, MIT Press, Cambridge, Mass.

25 Richard Byrne and Andrew Whiten (1985), 'Tactical deception of familiar individuals in baboons (*Papio ursinus*)', *Animal Behaviour* 33, 669–73; (1988), *Machiavellian Intelligence: Social expertise and the evolution of intellect in monkeys, apes and humans*, Clarendon, Oxford.

26 Chris Knight (1991), *Blood Relations: Menstruation and the origins of culture*, Yale University Press, New Haven.

27 Daniel C. Dennett (1995), *Darwin's Dangerous Idea: Evolution and the meanings of life*, Allen Lane, London.

28 Birgitta Sillén-Tullberg and Anders P. Møller (1993), 'The relationship between concealed ovulation and mating systems in anthropoid primates: a phylogenic analysis', *American Naturalist* 141, 1–25.

29 Nancy Knowlton (1979), 'Reproductive synchrony, parental investment and the evolutionary dynamics of sexual selection', *Animal Behaviour* 27, 1022–33.

30 Paul W. Turke (1984), 'Effects of ovulatory concealment and synchrony on protohominid mating systems and parental roles', *Ethology & Sociobiology* 5, 33–44.

31 Martha K. McClintock (1971), 'Menstrual synchrony and suppression', *Nature* 229, 244–45;
    http://www.mum.org/mensy71a.htm

32 Leonard Weller and Aron Weller (1993), 'Human menstrual synchrony: a critical assessment', *Neuroscience and Biobehavioral Reviews* 17, 427–39.

33 Leonard Weller and Aron Weller (1992), 'Menstrual synchrony in female couples', *Psychoneuroendocrinology* 17, 171–7.

34 Aron Weller and Leonard Weller (1997), 'Menstrual synchrony under optimal conditions: Bedouin families', *Journal of Comparative Psychology* 111(2), 143–151; (1998), 'Prolonged and very intensive contact may not be conducive to menstrual synchrony', *Psychoneuroendocrinology* 23(1), 19–32.

35 Kathleen Stern and Martha K. McClintock (1998), 'Regulation of ovulation by human pheromones', *Nature* 392, 177.

36 Natalie Angier, 'Study finds signs of elusive pheromones in humans', *New York Times* 12 March 1998.

37 Camilla Power and Leslie Aiello (1997), 'Female proto-symbolic strategies', in Lori D. Hager (ed.), *Women in Human Evolution*, Routledge, London.

38 Ibid; Chris Knight, op. cit.; Chris Knight, Camilla Power and Ian Watts, op. cit.

39 Ibid.

40 Matt Fraser, PaleoAnthro Lists Home Page,
    http://www.pitt.edu/~mattf/PalAntList.html

41 Robbie E. Davis-Floyd (1995), *Journal of the Royal Anthropological Institute* 1(1), 192–3.
42 R. A. Foley and C. M. Fitzgerald (1996), 'Is reproductive synchrony an evolutionarily stable strategy for hunter-gatherers?' *Current Anthropology* 37(3), 539–45.
43 Camilla Power, Catherine Arthur, and Leslie C. Aiello (1997), 'On seasonal reproductive synchrony as an evolutionarily stable strategy in human evolution', *Current Anthropology* 38(1), 88–91.
44 Robert C. Bailey et al. (1992), 'The ecology of birth seasonality among agriculturalists in Central Africa', *Journal of Biosocial Science* 24, 393–412; F. H. Bronson (1995), 'Seasonal variation in human reproduction: environmental factors', *Quarterly Review of Biology* 70(2), 141–64.
45 Robert Foley, personal communication.
46 Chris Knight, Camilla Power and Ian Watts, op. cit.
47 Ibid.
48 Ibid.; Ian Watts (1999), 'The origin of symbolic culture', in Robin Dunbar, Chris Knight and Camilla Power (eds), *The Evolution of Culture*, Edinburgh University Press, Edinburgh.
49 Robert Foley, 'Do we need a theory of human evolution?' (lecture), London School of Economics, 14 March 1996.
50 Chris Knight (1998), 'Ritual/speech coevolution: a solution to the problem of deception', in James R. Hurford, Michael Studdert-Kennedy and Chris Knight (eds), *Approaches to the Evolution of Language: Social and cognitive bases*, Cambridge University Press, Cambridge.
51 Robin Dunbar (1993), 'Co-evolution of neocortical size, group size and language in humans', *Behavioral and Brain Sciences* 16, 681–735; (1996), *Gossip, Grooming and the Evolution of Language*, Faber & Faber, London.
52 Camilla Power (1998), 'Old wives' tales: the gossip hypothesis and the reliability of cheap signals', in James R. Hurford, Michael Studdert-Kennedy and Chris Knight (eds), op. cit.
53 Steven Pinker (1994), *The Language Instinct: The new science of language and mind*, Penguin, London.
54 John R. Krebs and Richard Dawkins (1984), 'Animal signals: mind-reading and manipulation', in John R. Krebs and Nicholas B. Davies (eds), *Behavioural Ecology: An evolutionary approach*, 2nd edition, Blackwell, Oxford.

55 Chris Knight (1999), 'Sex and language as pretend-play', in Robin Dunbar, Chris Knight and Camilla Power (eds), op. cit.

Four Benefits

1 Richard Wrangham and Dale Peterson (1997), *Demonic Males: Apes and the origins of human violence*, Bloomsbury, London.

2 Quoted by Helena Cronin, 'Oh, those bonobos!', *New York Times Book Review* 29 August 1993 (review of Meredith F. Small, *Female Choices: Sexual behavior of female primates*, Cornell University Press, Ithaca).

3 Lee Berger (1998), 'Redrawing our family tree?' *National Geographic* 194(2), 90–9; Henry M. McHenry and Lee R. Berger (1998), 'Body proportions in *Australopithecus afarensis* and *A. africanus* and the origin of the genus *Homo*', *Journal of Human Evolution* 35, 1–22.

4 Henry M. McHenry (1996), 'Sexual dimorphism in fossil hominids and its socioecological implications', in James Steele and Stephen Shennan (eds), *The Archaeology of Human Ancestry: Power, Sex and Tradition*, Routledge, London.

5 Robert Foley (1996), 'The adaptive legacy of human evolution: a search for the environment of evolutionary adaptedness', *Evolutionary Anthropology* 4(6), 194–203; Robert Foley and Phyllis C. Lee (1989), 'Finite social space, evolutionary pathways and reconstructing hominid behavior', *Science* 243, 901–6.

6 Pierre L. van den Berghe (1986), 'Ethnicity and the sociobiology debate', in John Rex and David Mason (eds), *Theories of Race and Ethnic Relations*, Cambridge University Press, Cambridge.

7 Matt Ridley (1996), *The Origins of Virtue*, Viking, London.

8 Geoffrey Miller and Peter Todd, 'Evolution of vocabulary size through runaway sexual selection: theory, data and simulations', 'The Evolution of Language', 2nd International Conference, London, 1998.

9 Steve Jones, 'Go milk a fruit bat!', *New York Review of Books* 17 July 1997.

10 Geoffrey F. Miller (1997), 'How mate choice shaped human nature: A review of sexual selection and human evolution', in C. Crawford and D. L. Krebs (eds), *Handbook of Evolutionary Psychology: Ideas, issues and applications* 87–129, Lawrence Erlbaum, Mahwah, New Jersey;
http://ada.econ.ucl.ac.uk/papers/sex.htm

11 Kingsley Browne (1998), *Divided Labours: An evolutionary view of women at work*, Weidenfeld & Nicolson, London.

12 Stephen Jay Gould, 'Evolution: the pleasures of pluralism', *New York Review of Books* 26 June 1997.

13 R. Robin Baker and Mark A. Bellis (1993), 'Human sperm competition: ejaculate manipulation by females and a function for the female orgasm', *Animal Behaviour* 46, 887–909.

14 Kenan Malik, 'The Darwinian fallacy', *Prospect* December 1998.

15 Daniel C. Dennett (1995), *Darwin's Dangerous Idea: Evolution and the meanings of life*, Allen Lane, London.

16 Leda Cosmides and John Tooby (1997), 'Evolutionary Psychology: A Primer', Center for Evolutionary Psychology, University of California Santa Barbara,
    http://www.psych.uesb.edu/research/cep/primer.htm;
    (1992), 'Cognitive adaptations for social exchange', in Jerome H. Barkow, Leda Cosmides and John Tooby (eds) (1983), *The Modularity of Mind: An essay on faculty psychology*, MIT, Cambridge, Mass.

17 Jerry Fodor, op. cit.

18 Polly Toynbee, 'Birds do it, Bill does it', *The Guardian* 16 September 1998.

19 Richard Dawkins (1998), *Unweaving the Rainbow: Science, delusion and the appetite for wonder*, Penguin, London.

20 Robert Trivers, 'The logic of self-deception' (lecture), London School of Economics, 9 December 1998; W. D. Hamilton and M. Zuk (1982), 'Heritable true fitness and bright birds: a role for parasites?' *Science* 218, 384–7.

21 Leda Cosmides and John Tooby (1997), op. cit.

22 Robert Wright, 'The evolution of despair', *Time* 28 August 1995; Kenan Malik, 1998, op. cit.

23 The 'Unabomber Manifesto' is reproduced on several Web sites, e.g. http://www.thecourier.com/manifest.htm

24 William Irons (1998), 'Adaptively Relevant Environments versus the Environment of Evolutionary Adaptedness', *Evolutionary Anthropology* 6(6), 194–204.

25 Jerome H. Barkow, Leda Cosmides and John Tooby (eds) (1992), *The Adapted Mind: Evolutionary psychology and the generation of culture*, Oxford University Press, New York.

26 Robert Foley (1996), op. cit.

27 Martin Daly and Margo Wilson (1988), op. cit.

28 Robert Wright (1995), op. cit; (1994) *The Moral Animal: Evolutionary psychology and everyday life*, Pantheon, New York.

29 Martin Daly and Margo Wilson (1988), op. cit.

30 South African Police Services (1998), http://www.saps.co.za/8_crimeinfo/398/map1.htm; Ricky Taylor, (1998), 'Forty years of crime and criminal justice statistics, 1958 to 1997', Research Development Statistics, http://www.homeoffice.gov.uk/rsd/pdfs/40years.pdf

31 Robert Wright (1994), op. cit.

32 Randolph M. Nesse and George C. Williams (1995), *Evolution and Healing: The new science of Darwinian medicine*, Weidenfeld & Nicolson, London.

33 Timothy F. Doran et al. (1989), 'Acetaminophen: more harm than good for chickenpox?' *Journal of Pediatrics* 114(6), 1045–8.

34 Kenan Malik, 'Why illness means health', *Independent on Sunday* 23 June 1996.

35 Robert Wright (1995), op. cit.

36 Christopher J. L. Murray and Alan D. Lopez (1997), 'Global mortality, disability, and the contribution of risk factors: Global Burden of Disease Study', *Lancet* 349, 1436–42; 'Alternative projections of mortality and disability by cause 1990-2020: Global Burden of Disease Study', *Lancet* 349, 1498–1504.

37 Aisling Irwin, 'Shyness pill "to treat a serious problem"', *Daily Telegraph* 9 October 1998; Murray B. Stein et al. (1998), 'Paroxetine treatment of generalized social phobia (social anxiety disorder)', *Journal of the American Medical Association* 280, 708–713; Social Anxiety Disorder FAQ, American Psychiatric Association, http://www.degnanco.com/anxiety/faq.html

38 Randolph M. Nesse, 'The origins of suffering and the descent of medicine: when is it wise to block mental pain?' (lecture), London School of Economics, 3 July 1997.

39 John Maynard Smith (1998), *Shaping Life: Genes, embryos and evolution*, Weidenfeld & Nicolson, London; comments at launch for this title, London School of Economics, 8 October 1998.

40 Robert Trivers, op. cit.

41 Barbara Smuts (1995), 'The evolutionary origins of patriarchy', *Human Nature* 6(1), 1–32.

42 Robert H. Frank (1988), *Passions Within Reason: The strategic role of the emotions*, W. W. Norton, London.

43 Amotz and Avishag Zahavi (1997), *The Handicap Principle: A missing piece of Darwin's puzzle*, Oxford University Press, Oxford.

44 Robert H. Frank and Philip J. Cook (1995), *The Winner-Take-All Society*, Free Press, New York; Christian Tyler, 'Three tenors and a phenomenon', *Financial Times* 29 June 1996; Marek Kohn, 'Drive an Escort, not a Ferrari, and be happy', *Independent on Sunday* 16 June 1996.

45 David Erdal and Andrew Whiten (1994), 'On human egalitarianism: an evolutionary product of Machiavellian status escalation?' *Current Anthropology* 35, 175–83.

46 Christopher Boehm (1997), 'Egalitarian behaviour and the evolution of political intelligence', in Andrew Whiten and Richard W. Byrne (eds), *Machiavellian Intelligence II: Extensions and evaluations*, Cambridge University Press, Cambridge.

47 Richard G. Wilkinson (1996), *Unhealthy Societies: The afflictions of inequality*, Routledge, London.

48 Richard Wilkinson (1999), 'The culture of inequality', in *Income Inequality and Health: A reader*, I. Kawachi, B. P. Kennedy and R. G. Wilkinson (eds), New Press, New York.

49 Peter Singer (1999), *A Darwinian Left?* Weidenfeld & Nicolson, London.

50 Edward O. Wilson (1998), *Consilience: The unity of knowledge*, Little, Brown, London; (1975), *Sociobiology: The new synthesis*, Harvard University Press, Cambridge, Mass.

Further Reading

A selection of further reading; chosen more to stimulate than to summarize:

The Adapted Mind: Evolutionary psychology and the generation of culture, edited by Jerome H. Barkow, Leda Cosmides and John Tooby (1992), Oxford University Press, New York. This volume of papers sets out the principal themes of evolutionary psychology in the 1990s. 'The psychological foundations of culture', by Cosmides and Tooby, is both manifesto and *tour d'horizon*. Cosmides and Tooby are also the authors of 'Evolutionary Psychology: A Primer', published by the Center for Evolutionary Psychology on the University of California Santa Barbara Web site, http://www.psych.ucsb.edu/research/cep/primer.htm.

John Brockman (1995), *The Third Culture: Beyond the scientific revolution*, Simon & Schuster, New York. Scientific theories in conversational language, and what scientists say about each other.

Martin Daly and Margo Wilson (1998), *The Truth About Cinderella*, Phoenix, London. About relationships that are often fraught; between children and step-parents, between social science and Darwinism.

Daniel C. Dennett (1995), *Darwin's Dangerous Idea*, Allen Lane, London. Essential reading for anybody who has picked up the idea that modern Darwinism is little but zealotry and just-so stories. Dennett argues that resistance to the implications of Darwinism is not confined to Creationists, and can be especially dogged where the human mind is concerned.

Robert Foley (1995), *Humans Before Humanity*, Blackwell, Oxford. Foley's healthy emphasis on ecology produces an accessible account of humans in their natural habitats.

William Golding (1955), *The Inheritors*, Faber & Faber, London. The finest fictional imagination of minds that are human but different from ours, Golding's novel construes Neanderthals as prelapsarian innocents, and implies that what makes us human as we know it is original sin.

Women in Human Evolution, edited by Lori D. Hager (1997), Routledge, London. Sharp and readable chapters on issues such as the 'Palaeolithic glass ceiling'.

Annette Karmiloff-Smith (1992), *Beyond Modularity: A developmental perspective on cognitive science*, MIT Press, Cambridge, Mass. Squaring Piaget, evolutionary psychology and current thinking about children's cognitive development.

Geoffrey F. Miller (1997), 'How mate choice shaped human nature: A review of sexual selection and human evolution' in *Handbook of Evolutionary Psychology : Ideas, issues and applications*, edited by C. Crawford and D. L. Krebs, pp. 87–129, Lawrence Erlbaum, Mahwah, New Jersey; http://ada.econ.ucl.ac.uk/papers/sex.htm Miller (2000) discusses his ideas for a general audience in *The Mating Mind: How sexual choice shaped the evolution of human nature*, William Heinemann, London.

Matt Ridley (1993), *The Red Queen: Sex and the evolution of human nature*, Viking, London. The frequency with which this book is cited in scientific papers testifies to its author's grasp of the literature; his exuberant libertarian conservatism illustrates why sociobiology shouldn't be left to the Right.

Machiavellian Intelligence II: Extensions and evaluations, edited by Andrew Whiten and Richard W. Byrne (1997), Cambridge University Press, Cambridge. This collection shows what a rich and absorbing body of knowledge has grown from the insight that intelligence is a social issue.

Finally, the three books that influenced *As We Know It* the most:

Chris Knight (1991), *Blood Relations: Menstruation and the origins of culture*, Yale University Press, New Haven; especially the introduction.

Steven Mithen (1996), *The Prehistory of the Mind: A search for the origins of art, religion and science*, Thames & Hudson, London.

Amotz and Avishag Zahavi (1997), *The Handicap Principle: A missing piece of Darwin's puzzle*, Oxford University Press, Oxford.

Index

Ache people, 102

Acheulean tradition: development of biface form, 137, 155; ending, 130, 150–1; female roles, 130, 133; handaxes, 56–61, 98–9, 137, 139, 141, 154–5; learning, 133; origins, 92, 94, 98, 107; progress 57, 93, 158; territory, 76, 93, 94–5, 144; time-span, 93; toolmakers, 63, 68, 95, 125, 139–40, 144, 148–9, 165–6, 187

Adams, Douglas, 255

Adaptation and Natural Selection (Williams), 14

Adapted Mind, The (Barkow, Cosmides and Tooby), 260

Aesop, 187

Afar, 74

African-American communities, 22, 237

African Genesis (Ardrey), 86

Aiello, Leslie: on body energy budget, 104, 107; on female size, 128–9; on menstruation, 205, 211; on skulls, 106

algae, 36–7, 38

Altamira cave paintings, 173, 176

altruism, 277–9

Ancestral Passions (Morrell), 154

Andersson, Malte, 143

Animal Liberation (Singer), 290–1

Arcy-sur-Cure, 172, 174, 179–80, 185

Ardipithecus, 74

Ardrey, Robert, 85, 126, 233

Arthur, Catherine, 211

Atapuerca, Spain, 77, 95

Attachment and Loss (Bowlby), 11

Auden, W.H., 32

Auditorium Cave, India, 171–2

Aurignacian artifacts, 180

Australian Aboriginals, 208, 218

australopithecines: body form, 74, 78, 81, 85, 88–9, 128; brain size, 74, 78, 106, 108; development, 74–5; elf comparison, 81–2, 87; females, 81, 83, 85, 128; food supply, 83; groups, 81–3, 86; hominine distinction, 50; language question, 188, 190; mating system, 85–6; oldest remains, 43, 74; posture, 74, 86; taxonomy, 78, 79, 88–9; territory, 75, 83, 86; tooth size,

86; views of, 86–7
Australopithecus afarensis, 74, 75, 230
Australopithecus africanus, 74, 230
Australopithecus anamensis, 74
Australopithecus bahrelghazalia, 75
Australopithecus ramidus, 74
Axelrod, Robert, 14, 280
Ayoreo people, 262

babblers, 117–18, 120
baboons, 81, 193–4, 287
Baker, Robin, 245
Barker, Anna, 155
Barkow, Jerome, 260
Baron-Cohen, Simon, 13
beads, 179, 180, 182, 207
Bednarik, Robert, 175, 176, 177
Bellis, Mark, 245
Berekhat Ram, Golan Heights, 171, 176, 221–2
Berger, Lee, 230
Bergstrom, Carl, 119
Beyond Modularity (Karmiloff-Smith), 161–2
Binford, Lewis, 112–13
bipedalism, 80–1, 107
Black Panthers, 274
Blood Relations (Knight), 197, 198–9, 208
Boehm, Christopher, 283–4, 288
Boesch, Christophe, 132–3
bonobos: female bonding, 85, 229, 230; group life, 84, 85, 147, 228–30, 245; rules, 147; sex life, 84, 85, 228–9, 245; symbol systems, 92, 175, 234; toolmaking, 91–2; *see also* chimpanzees
Bowlby, John, 11, 259
Boxgrove, West Sussex: butchery, 96, 100, 112, 113, 136; diet, 100; handaxe production, 64–7,

96, 112, 127, 156; handaxe style, 139; handaxe use, 145; *Homo heidelbergensis*, 71–2, 79; horse butchery site, 112, 113, 115, 127, 136; human remains, 71–2, 77, 112; origins of handaxes, 63, 95; purpose of handaxes, 62; site, 70–1; social activities, 70, 115, 130, 186; spears, 113; toolmaking, 58–9; tools, 71, 115
Boxgrove Project, 96
Bradley, Bruce, 61, 144
brain size: australopithecine, 74, 78, 106, 108; diet and, 108–9; early hominid, 163–4; energy consumption, 103–5; *Homo* development, 78, 90–1, 106–8, 164, 216; increase in, 216, 248; Neanderthal, 78, 182, 191, 222; toolmaking and, 155; use of, 248
British Museum, 52, 68
Broca, Paul, 188
Broca's area, 188
Brockman, John, 34
Browne, Kingsley, 242, 252–3
Brunel-Deschamps, Eliette, 172
Burgess Shale, 255, 256
burials, 116, 182–3, 222
butchery, 110–16, 142, 165
Byrne, Richard, 193–4

Caddington handaxes, 61
Cain, 126
calendar, 176
Calvin, William H., 60, 97
Cambridge Encyclopedia of Human Evolution, 43
Carey, Susan, 167
Carter, George, 53
Cartmill, Matt, 190
Catholic Church, 8–9

charity, 278–9
Châtelperronian era, 174, 179, 180, 182, 185
Chauvet, Jean-Marie, 172
cheating, 145–6, 193–5, 211, 215
Cheers (TV series), 264
Cheney, Dorothy, 192
child development, 157–8, 162
childbirth, 261–3
chimpanzees: art, 175; classification, 73; demonism of, 228, 229; Gombe observations, 126–7, 154, 228; group sense, 235, 283; hunting, 101–2, 126–7, 130–1, 154; hypoglossal canal, 190; lowbrow hominids, 79; male kin-bonding, 84, 233–4; mating system, 82, 85, 102, 130–1, 154; sexual differences, 85; symbol use, 194; teaching skills, 132–3, 165; tool use, 132–3, 165, 215, 234; *see also* bonobos
China, 76
Chomsky, Noam, 160–1
Christianity, 247, 254
church architecture, 163–7
Ciochon, Russell L., 93
Clactonian assemblages, 95, 144
climate, 80, 83, 93, 137
Clinton, Bill, 251, 253
clothing, 70
Clottes, Jean, 177–8
Clutton-Brock, Tim, 108
coalition, 232–5
Cockburn, Alexander, 103
Conrad, Joseph, 233
consilience, 293
Constantinople, 234, 235
Cook, Philip J., 280–2
cooking, 207
cosmetics, 205
Cosmides, Leda: on evolutionary

psychology, 260, 264; on human mind, 257; on modularity, 161, 249; on social contract tests, 250; on SSSM, 19
Cougnac caves, 25–6
Cronin, Helena, 13

Daly, Martin, 19, 21–2, 23–4, 261–2
Dart, Raymond, 86, 87, 126
Darwin, Charles: influence, 22, 28, 241, 242, 288–9; Law of Equal Inheritance, 131; Mendel's work, 34; natural selection theory, 38, 120, 198; on bipedality, 107; on tooth size, 86; sexual selection concept, 62, 120
Darwin Seminars, 13, 20, 139
Darwinian Left, A (Singer), 291
Darwinism Today, 242
Darwin's Dangerous Idea (Dennett), 8, 198
Davidson, Iain, 61, 63–4, 97
Dawkins, Marian Stamp, 123–4
Dawkins, Richard: materialism, 199; on Argument from Personal Incredulity, 153; on mass extinctions, 255; on memes, 247; on phenotypes, 138; on selfish gene, 14; on signalling, 218
Dawson, Peter, 96, 97, 99
deer, 69, 96, 165
Demonic Males (Wrangham and Peterson), 228
Demos think-tank, 41
Dennett, Daniel: materialism, 199; on Darwinian selection, 198; on 'good trick', 164; on Margulis' work, 35; on memes, 248; on Social Darwinism, 8

depression, 269–70, 287
d'Errico, Francesco, 179, 181
Descartes, René, 38
Descent of Man, The (Darwin),
    131–2
Dezzani, Ray, 59
Diamond, Jared, 73, 101, 181
disability-adjusted life years
    (DALYs), 269
Dmanisi, Georgia, 77
DNA, 44, 45
Donald, Merlin, 192, 196
drama, 220
Dunbar, Robin, 212–13, 216–17
Durkheim, Émile, 22, 241

Efe people, 212
Eland Bull Dance, 208, 218
elves, 81–2, 87
emotions, 267–8
Enfield, Harry, 278
Engels, Friedrich, 197
environment of evolutionary
    adaptation (EEA), 11, 257,
    259, 263
Erdal, David, 283
Ethiopia, 42, 54, 58, 73–4, 81,
    89
evolutionary psychology, 10–12
exaptation, 189

Fairweather Eden (Pitts and
    Roberts), 96
Falk, Dean, 107, 129, 188
family size, 82–3
female: australopithecines, 81, 83,
    85, 128; body, 15–18, 128; body
    size, 128–9; bonobos, 85, 229,
    230; handaxe production,
    131–4; menstrual synchrony,
    200–4, 211–13, 222;
    menstruation, 204–7; roles,
    100–1, 111, 128, 130–4

feminism, 274–5, 275–6
ferric oxide, 184
fever, 267, 271
Fisher, Ronald, 117, 143
Fitzgerald, C.M., 210, 211
flint, 51–2, 57, 65–8
Flint Implements (British
    Museum), 52, 56
flying, 266
Fodor, Jerry, 159–62, 250
Foley, Robert: on
    australopithecine social
    structure, 84; on bipedalism,
    80; on female social bonds,
    231; on food supply, 83; on
    human evolution, 180, 214; on
    hunter-gatherers, 260; on meat
    consumption, 103; on primate
    social space, 232; on
    reproductive strategies, 82, 85,
    130, 150–1; on synchrony, 210,
    211, 212
food: diet, 100–3; gathering,
    100–1; supply, 83
Franco-Cantabrian cave-art, 176
Frank, Robert H., 276–82
Freeman, Derek, 228
Freud, Sigmund, 5, 157, 242,
    254
Furze Platt, Berkshire, 51, 53
Furze Platt Giant, 53–4, 135

Gaia hypothesis, 35, 36
Gamble, Clive, 70, 115, 191
gazelles, 118
Gibson, Kathleen, 108
Golan Heights, 171, 176
Golding, William, 183
Gombe, Tanzania, 126, 154, 228
Gona, Ethiopia, 89, 91, 92
Goodall, Jane, 126–7, 130
Goodwin, Brian, 36, 177
Goodwin, Frederick, 22

gorillas: classification, 73; diet,
     229; mating arrangements, 82;
     sexual differences, 81; social
     system, 82, 85
Gould, Stephen Jay: on
     evolutionary theory, 27, 36,
     243, 255–6; on exaptation, 189;
     on language, 161; on ultra-
     Darwinism, 199, 255
Gowlett, John, 57, 158
Grafen, Alan, 119–20, 139, 143
Gran Dolina, Spain, 77
Great Leap Forwards, 181
Grimes Graves, 65
grooming, 217
Grotte Chauvet, 173, 178
Grotte du Taï, 176
grouse, 123–4
Guilford, Tim, 123–4

Hadar, Ethiopia, 81
Hadza people, 207–8, 211
Haig, David, 13, 20–1
Haldane, J.B.S., 232
Halverson, John, 178
Hamilton, W.D., 257
Hamilton, William, 14, 37, 276
hammers, antler, 71, 115
handaxes: Boxgrove site, 70, 71,
     95, 96, 100, 115, 145; butchery
     tool, 59, 70, 96, 99–100, 112,
     115, 122, 142, 154; cheating
     question, 145–6; date range,
     54, 77, 94, 153; débitage, 65, 68,
     71, 112, 156; distribution, 54,
     77, 94–5, 153; flint, 51–2, 57,
     67; Furze Platt industry, 51, 53;
     landscape factor, 144–5;
     Levallois technique, 64;
     materials, 51–2, 57, 156;
     missiles, 59–60; modern
     production, 59, 61, 64–8, 96–7,
     124–5; origins, 57–8, 63–4,

92–3, 94–5, 98–9, 107, 136–7;
     purpose, 56–7, 63, 122–3,
     138–9; risks in production, 65,
     126; sexual selection theory,
     62–3, 123–6, 134–6, 137–42,
     143–4, 145, 148–9, 151, 152–6;
     shapes, 54–5, 57, 58–9, 97–8;
     size, 53–4, 59, 98–9; study of,
     5–6, 79; symmetry, 59, 61, 62,
     137, 148, 152, 155; transport of,
     98–9; uses, 59–60, 96–8
Handicap Principle: application,
     120; babies' signalling, 120–1;
     charity, 278–9; development,
     117–18; handaxe knapping,
     124, 138–40; influence, 143;
     mathematical models, 33, 118,
     143; publication, 33, 118, 143;
     sexual selection, 117, 121;
     types of handicap, 119–20
Haraway, Donna, 100, 199
Harding, Phil, 64–7, 96, 124, 140
Harvey, Paul, 108
Hauser, Marc, 192
Hawking, Stephen, 9
Hayden, Brian, 64
Hegel, Georg Wilhelm Friedrich,
     164
Hillaire, Christian, 172–3
Hinde, Robert, 16
Hitch-hiker's Guide to the Galaxy,
     The (Adams), 255
Hohlenstein-Stadel, Germany,
     167
Holloway, Ralph, 107
Homicide (Daly and Wilson),
     19–20
hominines, 50, 75, 230
Homo: brain size, 90–1, 103, 106,
     108; Broca's area, 188;
     classification, 50, 79; diet,
     89–90; early remains, 89;
     evolution, 91; variety, 50

Homo antecessor, 77

Homo erectus: body form, 91; body size, 230; classification, 76, 79, 80; diet, 231; handaxes, 154; larynx, 189; mind, 164; Ngangdong remains, 43, 44; origins, 91

Homo ergaster: body form, 91; body size, 230; brain size, 78, 106; classification, 76, 79, 80; disappearance, 93; habitat, 231; handaxes, 93, 136–7; mind, 164; origins, 91; toolmaking, 136–7

Homo habilis, 75, 79, 91, 106, 190

Homo heidelbergensis, 71–2, 77, 79, 93, 190

Homo rhodesiensis, 190

Homo rudolfensis, 79, 106

Homo sapiens: archaic, 71–2, 77, 106; brain size, 78, 216; classification, 50, 76, 79–80; establishment, 43, 216; hypoglossal canal, 190; larynx, 189; tools, 50

Homo sapiens heidelbergensis, 79

Horgan, John, 16

horses: butchery, 112, 126, 127, 136; groups, 187; hunting, 69, 113, 114, 165

How the Mind Works (Pinker), 246, 259

Human Behavior and Evolution Society, 11, 16

Humphrey, Nicholas, 109

hunting, 101, 113, 115, 131; chimpanzee, 126–7

hyenas, 113, 115

hyoid bone, 190–1

infant mortality, 210–11, 268–9

infanticide, 261–3, 268

Inheritors, The (Golding), 183

initiation rites, 218

instincts, 248–9

intelligence: general, 165, 166–7, 239; natural history, 166; social, 165–6, 167; technical, 165–6, 167

Inuit people, 113

IQ tests, 239

Irons, William, 259, 261, 262–3, 269

Isaac, Glynn, 57, 59

Isimila, 98

James, William, 249

Japan, 286

Java, 43, 76, 91

John Paul II, Pope, 8, 9, 153

Jolly, Alison, 109

Jones, Peter, 97–8, 99, 156

Jones, Steve, 125, 240

Just-So Stories (Kipling), 26–32

Kaczynski, Theodore, 258

Kamin, Leon, 10

Kanzi (bonobo), 91, 158–9

Karmiloff-Smith, Annette, 161–2, 167

Kimunjia, Mzalendo, 92

King, Barbara, 147

kinship, 236–7

Kipling, Rudyard, 27–32

knapping, 65–7, 91–2, 124–6, 146–7

Knauft, Bruce, 283

Knight, Chris: on cheating, 215; on early culture, 198–9, 207, 210, 212, 241; on Rainbow Snake, 208–9; on ritual, 199, 218, 219, 220; on sociobiology, 197, 199; on trust, 196; on vervet monkeys, 194; politics, 197, 274

Knowlton, Nancy, 200

Konso-Gardula, Ethiopia, 54
Krebs, John, 218
Kubrick, Stanley, 100
!Kung people, *see* Zu/'hoasi

La Ferrassie, France, 182
Labour Party, 40
Laetoli, Tanzania, 74
Lahr, Marta, 180
Laitman, Jeffrey, 189
Lake, Mark, 164
Lamarck, Jean de, 29–30
language: alarm systems, 192–3,
    195–6; Boxgrove hominids,
    115, 191–2; chimpanzees,
    194–5; Chomsky's work,
    160–1; development, 108,
    188–95, 215, 220; mime, 192,
    196, 219; Neanderthals, 190–1;
    physical evidence, 188–91,
    195; thought and, 167–8; *see
    also* speech
Larick, Roy, 93
larynx, 189
Lascaux cave paintings, 173, 176
Latino communities, 21–2
Law of Equal Inheritance, 131
Leakey, Louis, 58, 59
Leakey, Mary, 58, 74, 89, 154
Leakey, Richard, 50, 76, 96, 97
Lee, Phyllis: on australopithecine
    social systems, 83–4; on female
    social bonds, 231; on meat
    consumption, 103; on primate
    social space, 232; on
    reproductive strategies, 82, 85,
    130, 151; on social systems, 84
lemurs, 109
Lenton, Tim, 37
Leroi, Armand Marie, 138–9
Leroi-Gourhan, André, 172, 176,
    178
Lese people, 212

Levallois technique, 64
Lévi-Strauss, Claude, 197
Lewin, Roger, 96, 97
Lewis-Williams, David, 176,
    177–8
Lewontin, Richard, 10, 35
Lieberman, Philip, 189
lions, 113
Lippman, Abby, 20–1
Lokalalei, Lake Turkana, 92
Lord, John, 124–5, 126, 135, 140,
    155
Lovejoy, Owen, 123, 129
Lovelock, James, 35, 36
Lower Palaeolithic: art, 171–2;
    Boxgrove site, 49–50, 70, 71,
    112, 186; division from Middle
    Palaeolithic, 180; handaxes,
    5–6, 51, 141, 145; hominid
    classification, 50; hominid life,
    5–6, 186; human remains, 71;
    spears, 114; tool industries,
    174–5
Lucy (hominid), 43, 74

macaque monkeys, 132
McClintock, Martha, 202–3
McHenry, Henry M., 85, 86, 128,
    230–1
Malik, Kenan, 40
mammoths, 44
manganese dioxide, 184, 185
Margulis, Lynn, 34–5, 36, 37
Marshack, Alexander, 171, 176,
    180, 182
Marx, Karl, 197, 242, 254
masturbation, 244–5
Matata (bonobo), 92
maternal bonding, 261–2
Matsigenka people, 18
Mauer remains, 72, 77
Mead, Margaret, 228
meat consumption, 100–3, 207

memes, 247–8
Mendel, Gregor, 28, 34
menstruation, 204–7; synchrony,
    200–4, 211, 222
Middle Palaeolithic: art, 172, 178;
    burials, 116; tools, 174, 175,
    178, 180
Middle Pleistocene, 57
Miller, Geoffrey, 41, 121, 134,
    139, 239–42
mime, 192, 196, 219
mind: architecture, 51, 163–7;
    modular scheme, 159–62;
    Piaget's model, 157–9, 162
Mitchell, John, 96, 97, 99
Mithen, Steven: flint-knapping
    workshop, 124, 135; on
    cheating, 145; on handaxe
    symmetry, 144, 152, 154, 157;
    on mind, 163–7, 181; on
    technical intelligence, 115;
    students, 155
mitochondria, 35
Modularity of Mind, The (Fodor),
    159
Møller, Anders Pape, 13, 85,
    199–200
monogamy, 81, 84, 85, 129, 130
moon, phases of, 206
Moral Animal, The (Wright), 264
Morrell, Virginia, 154
Morris, Desmond, 175, 176
Mortillet, Gabriel de, 56
Moulay Ismail the Bloodthirsty,
    82, 276
Mount Carmel, Israel, 190
Mousterian tradition, 178, 180,
    182, 185
M25, 33–4
Mulgan, Geoff, 40–1
music, 42

Nariokotome Boy, 76

nationalism, 236
Natural History Museum, South
    Kensington, 16, 53
Nature, 43, 114
Nazism, 254, 273
Neanderthals: ancestry, 77; art,
    172, 179; body form, 78, 128,
    179; brain size, 78, 182, 191,
    222; burials, 116, 182–3, 222;
    DNA, 44, 45; modern human
    comparison, 178–80, 222–3;
    music, 42; ochre use, 185;
    speech, 183, 189, 190–1;
    territory, 43; toolmaking, 64,
    175, 178, 182; way of life,
    178–80, 181–3
Nesse, Randolph M., 267–8, 269,
    270, 271
New Guinea, 268
New Statesman, 40
New York Times, 20
New Yorker, 251
Newcomer, Mark, 68
Newton, Huey, 14, 274, 277
Ngandong, Java, 43, 44
Noble, William, 61, 63–4

O'Brien, Eileen, 59, 97
obsidian, 53, 57
ochre, 184–5, 205, 207, 208, 214
Oldowan tradition: Developed
    Oldowan tools, 98; duration,
    89; end, 91; origins, 57–8, 89,
    94; relationship with
    Acheulean industry, 98; shape
    of tools, 58, 92–3, 137;
    toolmakers, 58, 89, 165;
    toolmaking skills, 158, 165
Olduvai Gorge, 49, 54, 57–8, 99,
    156
Olorgesailie, 60, 98–9, 100
On Growth and Form (Thompson),
    177

orang-utans, 82, 84
Orce, Spain, 77
Origin of Species, The (Darwin), 289
ovulation, 199–200, 211

Pakistan, 76
Paleolithic Venuses, 16–17
Pan paniscus, 85, 228–30, 234
Pan troglodytes, 85, 228, 229–30,
    234
paranthropines, 75, 79, 89–90
Paranthropus boisei, 89
Parker, Sue, 108
Passions Within Reason (Frank),
    276–7
patterns, 177
peacocks, 117
Perception, 178
Peterson, Dale, 228, 230
Piaget, Jean, 157–8, 159, 162
Pinker, Steve, 246, 251, 253,
    259
Pitts, Michael, 96
Pius XII, Pope, 8
Plato, 219
Plavcan, J. Michael, 86, 128
Poles, 237
polygyny, 85
postmodernism, 40
Power, Camilla, 128–9, 204–5,
    211–12, 217
Prehistory of the Mind, The
    (Mithen), 163
Primate Visions (Haraway), 100,
    199
Prozac, 270

Qafzeh, Israel, 167
quartzite, 53, 57
Quneitra, Golan Heights, 176,
    178

Rainbow Snake, 208–9, 218

religion, 219, 242, 247
reproductive strategies, 81–6
rhinoceroses, 69, 114, 178
Ricci (chimpanzee), 132–3
Ridley, Matt, 233, 234
*Rise and Fall of the Third
    Chimpanzee, The* (Diamond),
    101
risk, 251–2
ritual, 124, 199, 207, 218–20
Roberts, Mark, 70, 96, 112, 114,
    115
Rose, Steven, 10
Roseto, Pennsylvania, 286
Rules, 146–7
Runciman, Garry, 14
Runciman, W.G., 138–9
Russians, 237

Sacks, Oliver, 177
Salomé (chimpanzee), 132
Samoa, 228–9, 268, 286, 288
Sampson, C. Garth, 61, 144
San people, 177, 207–8, 260
Sapolsky, Robert, 287
Savage-Rumbaugh, Sue, 91
Schick, Kathy, 59
Schöningen, Germany, 114,
    141
Seale, Bobby, 274
selective serotonin reuptake
    inhibitors (SSRIs), 269–71
Serbs, 237
sexual selection: concept, 62;
    female body, 128; genes or
    good taste, 143; handaxe-
    making competence, 62–3,
    133–7, 141, 144–9, 153–4, 155;
    Handicap Principle, 117, 121,
    143, 150; language and culture,
    239–40
sexuality, human, 245–6
Seyfarth, Robert, 192

shamanism, 176, 178
Shanidar, Iraq, 116
Shanker, Stuart, 147
Siegel, Ronald K., 177
Sillén-Tullberg, Birgitta, 85, 199–200
Singer, Peter, 290–1, 294
Singh, Devendra, 15–16, 18
Skhul, Israel, 167
skulls, 189–90, *see also* brain
sloths, giant, 44–5
Slovenia, 442
Small, Meredith, 229
Smith, Adam, 274
Smith, John Maynard: on co-operation and group selection, 14, 36; on evolutionarily stable strategy, 39; on Handicap Principle, 118, 119; on politics, 273
Smuts, Barbara, 274–5
Sober, Elliott, 36
Social Animal, The (Runciman), 138–9
Social Darwinism, 7–8, 10, 12, 139
social systems, 84
sociobiology, 9–10, 11–12, 197
Sociobiology: The New Synthesis (Wilson), 9, 293
spears, 42, 113, 114, 141
speech, 183, 188, 191, 217–20, *see also* language
Spelke, Elizabeth, 167
Sperber, Dan, 160, 250
Standard Social Science Model (SSSM), 19, 41
step-parents and step-children, 19–24, 38
Stringer, Chris, 43, 191
Swanscombe, 71
symmetry, 137, 144, 152
Symons, Donald, 10–11

synchrony, female, 200–4, 211–13, 222

Taï, Ivory Coast, 130, 132, 165
Tata, Hungary, 172, 176, 185
Thames Valley, 49–50, 53
Thatcher, Margaret, 241, 274
Theory of the Leisure Class (Veblen), 120, 278
Thieme, Hartmut, 114–15
Third Culture, The (Brockman), 34
Thompson, D'Arcy, 177
Tit for Tat, 279–80
Tobias, Phillip, 74, 188
Tomlin, Lily, 217
Tooby, John: on *Cheers*, 264; on EEA concept, 257, 260; on mind modularity, 161, 249; on SSSM, 19
tooth size, 86
Toth, Nicholas, 59
Toynbee, Polly, 253
Trivers, Robert: on co-operation, 15; on reciprocal altruism, 276; on scientific papers, 256–7; on self-deception, 14, 274, 278
Truth about Cinderella, The (Daly and Wilson), 21
Turkana, Lake, 49, 50, 92, 94
Turkana Boy, 76
Turke, Paul, 201, 211
Twiggy, 16
2001 (film), 100, 181

Ubeidiya, Ethiopia, 77
Unweaving the Rainbow (Dawkins), 255
Updike, John, 125
Upper Palaeolithic: art, 5, 172–3, 176, 177–8; calendar, 176; division from Middle Palaeolithic, 174, 180; toolmaking, 175

van den Berghe, Pierre, 232
van Schaik, Carel P., 86
Vasilyev, Mrs, 83
Veblen, Thorstein, 120, 278
Venuses, 16–18
vervet monkeys, 192–3, 194, 195
vocabulary, 239
Vrba, Elizabeth, 189

Wallacean islands, 43
Wamba, Zaïre, 147
warfare, 218
Wason test, 249, 250
Watts, Ian, 184, 214
Weller, Leonard and Aron, 202, 203
Wells, H.G., 59
Wheeler, Peter, 80, 104, 106, 107
White, Randall, 18, 175
Whitehall studies, 285
Whitehouse, Paul, 278
Whiten, Andrew, 193–4, 283
Why Men Don't Iron (TV series), 39
Wilkinson, Richard, 282, 284, 285–8
Willendorf Venus, 17

Williams, George, 14, 35–6, 267–8
Wilson, David Sloan, 36
Wilson, Edward O., 9, 12, 293
Wilson, Margo, 19, 21–2, 23–4, 261–2
Winner-Take-All Society, The (Frank and Cook), 280–2
wolves, 113, 186–7, 195
Wonderful Life (Gould), 255
Wood, Bernard, 80, 89
World Health Organisation (WHO), 269
Wrangham, Richard, 228, 229, 230
Wright, Robert, 264, 266
Wright, Sewall, 84
Wymer, John, 49–50, 51
Wynn, Thomas, 97, 157–9

xenophobia, 234, 235, 236

Zahavi, Amotz and Avishag, 32–3, 117–21, 138, 139–40, 143, 279
Zu/'hoasi people, 102–3, 208, 211, 212, 262, 265
Zuk, M., 257